Springer Theses

Recognizing Outstanding Ph.D. Research

AF587603

Aims and Scope

The series "Springer Theses" brings together a selection of the very best Ph.D. theses from around the world and across the physical sciences. Nominated and endorsed by two recognized specialists, each published volume has been selected for its scientific excellence and the high impact of its contents for the pertinent field of research. For greater accessibility to non-specialists, the published versions include an extended introduction, as well as a foreword by the student's supervisor explaining the special relevance of the work for the field. As a whole, the series will provide a valuable resource both for newcomers to the research fields described, and for other scientists seeking detailed background information on special questions. Finally, it provides an accredited documentation of the valuable contributions made by today's younger generation of scientists.

Theses are accepted into the series by invited nomination only and must fulfill all of the following criteria

- They must be written in good English.
- The topic should fall within the confines of Chemistry, Physics, Earth Sciences, Engineering and related interdisciplinary fields such as Materials, Nanoscience, Chemical Engineering, Complex Systems and Biophysics.
- The work reported in the thesis must represent a significant scientific advance.
- If the thesis includes previously published material, permission to reproduce this must be gained from the respective copyright holder.
- They must have been examined and passed during the 12 months prior to nomination.
- Each thesis should include a foreword by the supervisor outlining the significance of its content.
- The theses should have a clearly defined structure including an introduction accessible to scientists not expert in that particular field.

More information about this series at http://www.springer.com/series/8790

Daniel Kiefer

Relativistic Electron Mirrors

from High Intensity Laser–Nanofoil Interactions

Doctoral Thesis accepted by
Ludwig-Maximilians-University of Munich,
Garching, Germany

Springer

Author
Dr. Daniel Kiefer
Ludwig-Maximilians-University of Munich
Garching
Germany

Supervisor
Prof. Jörg Schreiber
Ludwig-Maximilians-University of Munich
Garching
Germany

ISSN 2190-5053 ISSN 2190-5061 (electronic)
ISBN 978-3-319-38391-0 ISBN 978-3-319-07752-9 (eBook)
DOI 10.1007/978-3-319-07752-9

Springer Cham Heidelberg New York Dordrecht London

© Springer International Publishing Switzerland 2015
Softcover reprint of the hardcover 1st edition 2015
This work is subject to copyright. All rights are reserved by the Publisher, whether the whole or part of the material is concerned, specifically the rights of translation, reprinting, reuse of illustrations, recitation, broadcasting, reproduction on microfilms or in any other physical way, and transmission or information storage and retrieval, electronic adaptation, computer software, or by similar or dissimilar methodology now known or hereafter developed. Exempted from this legal reservation are brief excerpts in connection with reviews or scholarly analysis or material supplied specifically for the purpose of being entered and executed on a computer system, for exclusive use by the purchaser of the work. Duplication of this publication or parts thereof is permitted only under the provisions of the Copyright Law of the Publisher's location, in its current version, and permission for use must always be obtained from Springer. Permissions for use may be obtained through RightsLink at the Copyright Clearance Center. Violations are liable to prosecution under the respective Copyright Law.
The use of general descriptive names, registered names, trademarks, service marks, etc. in this publication does not imply, even in the absence of a specific statement, that such names are exempt from the relevant protective laws and regulations and therefore free for general use.
While the advice and information in this book are believed to be true and accurate at the date of publication, neither the authors nor the editors nor the publisher can accept any legal responsibility for any errors or omissions that may be made. The publisher makes no warranty, express or implied, with respect to the material contained herein.

Printed on acid-free paper

Springer is part of Springer Science+Business Media (www.springer.com)

Supervisor's Foreword

One of the most fascinating consequences of Einstein's theory of special relativity is that light reflected from a mirror moving with velocities close to the speed of light is frequency upshifted. While the relativistic Doppler-effect is frequently used in incoherent brilliant light sources around the globe, generating a mirror structure, which can per definition reflect light coherently, has remained illusionary. The advent of femtosecond laser pulses with peak powers approaching 1 PW promised a viable route to create such mirrors. Numerous theoretical investigations had shown that relativistic electron sheets approaching solid densities can be generated when irradiating nanometer thin foils at intensities well beyond relativistic intensities of 10^{18} W/cm^2.

The ultimate goal of Daniel Kiefer's work was to realize those purely theoretical concepts in experiments with state-of-the-art high-power laser systems. One major difficulty has been the formation of the mirror itself, which relies on complex dynamics taking place when high-intensity laser pulses interact with nanometer thin plasmas. The acceleration of the dense electron sheets had been barely investigated. In addition, the peak intensity of the laser is about 10 orders of magnitude beyond typical damage thresholds of the thin-foil material. To ensure survival of the fragile target, the laser needs the accordingly high temporal contrast. Its intensity must spurt over 10 orders of magnitude within a few picoseconds, which poses a significant experimental challenge and typically requires the use of plasma mirrors. Last but not least, the mirror remains in its required properties, one of which being the high electron density, for a few femtoseconds only. This is a very short window during which a counter-propagating laser pulse can be reflected and frequency upshifted in order to observe Einstein's original idea.

Daniel started off measuring the properties of electrons accelerated from laser-irradiated nanometer thin foils at the most advanced high-power, high-contrast laser facilities around the world. He collected the most comprehensive data set, which covers laser energies from one to hundreds of Joules and pulse durations from a few tens to a few hundreds of femtoseconds. His measurements and observations, as well as his contribution to various campaigns, were vital and impacted on many levels, for example to improve the understanding of laser-driven ion acceleration and the generation of short radiation pulses. The interpretation and quantitative understanding of the electron data, however, remained difficult. Daniel invested a substantial period to gain deeper understanding using the most advanced particle-in-cell simulations, the standard numerical tool for describing the physics of laser–plasma interaction at relativistic intensities. His studies revealed what we had already suspected. The electron mirrors do form in realistic conditions, but they are more fragile as compared to therefore mentioned idealistic calculations. Moreover, their signature was hardly measurable by means of indirect tools such as electron spectrometers. Reflecting a counter-propagating laser pulse off the short-lived dense electron sheets reemerged as a unique way to gain knowledge about relativistic high-density plasma-physics.

Daniel Kiefer's central experiment was part of a campaign to study high-intensity laser-based XUV-plasma sources at Rutherford-Appleton-Laboratory's Central Laser Facility. The ASTRA Gemini laser features two synchronized laser pulses and high temporal contrast, two of the main requirements for his ambitious study. It is worth mentioning that using 100–200 nm thick foils resulted in bright XUV-emission, more specifically coherent synchrotron radiation (CSR), a discovery to which Daniel significantly contributed. This emission was not observed for target thicknesses below 50 nm. Instead, when sending the counter-propagating pulse with the exact (fs) timing to the main driving laser, Daniel observed a significant signal in the photon spectrometer. It was this simple result, which he was anxious about for years, as this signal was already the proof that the back-scattered radiation must have been generated in a highly coherent process, and could therefore be interpreted as a reflection. The wavelength of the reflected light was blue-shifted by a factor of 10, the spectrum was broad and modulated. Combined with his simulation results, Daniel concluded that electron sheets are ejected every 2.7 fs, the period of the driving laser, and consist of electrons with a broad range of relativistic energies. His fascinating observation is not only the first realization of Einstein's Gedanken-experiment. The characteristics of the reflected radiation allowed also valuable insights into the complex dynamics of lasers interacting with nanoscale plasmas at highest intensities. Most intriguing though seems the possibility of creating short laser pulses in the ultraviolet or even X-ray region, with pulse durations well below 1 fs.

Garching, Germany, April 2014 Prof. Jörg Schreiber

Abstract

The reflection of a laser pulse from a mirror moving close to the speed of light could, in principle, create an X-ray pulse with unprecedented high brightness owing to the increase in photon energy and accompanying temporal compression by a factor of $4\gamma^2$, where γ is the Lorentz factor of the mirror. While this scheme is theoretically intriguingly simple and was first discussed by A. Einstein more than a century ago, the generation of a relativistic structure, which acts as a mirror, is demanding in many different aspects. Recently, the interaction of a high-intensity laser pulse with a nanometer thin foil has raised great interest as it promises the creation of a dense, attosecond short, relativistic electron bunch capable of forming a mirror structure that scatters counter-propagating light coherently and shifts its frequency to higher photon energies. However, so far, this novel concept has been discussed only in theoretical studies using highly idealized interaction parameters.

This thesis investigates the generation of a relativistic electron mirror from a nanometer foil with current state-of-the-art high-intensity laser pulses and demonstrates for the first time the reflection from those structures in an experiment. To achieve this result, the electron acceleration from high-intensity laser nanometer foil interactions was studied in a series of experiments using three inherently different high-power laser systems and free-standing foils as thin as 3 nm. A drastic increase in the electron energies was observed when reducing the target thickness from the micrometer to the nanometer scale. Quasi-monoenergetic electron beams were measured for the first time from ultrathin ($\leq$5 nm) foils, reaching energies up to $\sim$35 MeV. The acceleration process was studied in simulations well-adapted to the experiments, indicating the transition from plasma to free electron dynamics as the target thickness is reduced to the few nanometer range. The experience gained from those studies allowed proceeding to the central goal, the demonstration of the relativistically flying mirror, which was achieved at the Astra Gemini dual beam laser facility. In this experiment, a frequency shift in the backscatter signal from the visible (800 nm) to the extreme ultraviolet ($\sim$60 nm) was observed when irradiating the interaction region with a counter-propagating probe pulse simultaneously. Complementary to the experimental observations, a detailed numerical study on the dual beam interaction is presented, explaining the mirror formation and reflection process in great depth, indicating a $> 10^4$ fold increase in the backscatter efficiency as compared to the expected incoherent signal. The simulations show that the created electron mirrors propagate freely at relativistic velocities while reflecting off the counter-propagating laser, thereby truly acting like the relativistic mirror first discussed in Einstein's thought experiment. The reported work gives an intriguing insight into the electron dynamics in high-intensity laser nanofoil interactions and constitutes a major step toward the coherent backscattering from a relativistic electron mirror of solid density, which could potentially generate bright bursts of X-rays on a micro-scale.

Acknowledgments

At this point, I would like to thank all of my colleagues, collaborators, and all staff members, who made this work happen. During my time as a Ph.D. student, I was lucky working with so many excellent people from all over the world, certainly making it an unforgettable experience. I especially would like to thank the following people:

- Prof. Jörg Schreiber who has been a great supervisor, helping me with many brilliant ideas, giving me as much freedom as I wanted and took care of me whenever it was needed. The countless hours of theory discussions I spent with him have been one of the most enjoyable part of my work, which I certainly do not want to miss.
- I would like to thank equally Prof. Dietrich Habs, a true visionary, who was able to infect me with his enthusiasm over and over again and who certainly made an everlasting impression on me. Not to forget, I would like to thank him deeply for providing me all resources including an unlimited travel budget, which made this work possible.
- I would like to thank Prof. Matthew Zepf, who inviting me to the Astra Gemini experiment and who was kindly willing to review my thesis last minute.
- I would like to thank Prof. Hartmut Ruhl for supporting my work, especially for hosting me in his theory group downtown for more than a year.
- I am grateful to Prof. Ferenc Krausz for giving me the opportunity to be part of his excellent group at the MPQ. I am also indebted to Prof. Manuel Hegelich, who always gave me a warm welcome in Los Alamos. I want to thank Prof. Jürgen Meyer-ter-Vehn for his persistent interest in my work. I am also very grateful to Prof. Toshiki Tajima, who certainly “accelerated” our research group and who shared many discussions with me at lunch time.
- My special thanks go to Sergey Rykovanov, who introduced me to the fabulous world of PIC simulations and who helped me advancing my physical understanding considerably. It has been an inspiration and a great pleasure to work with him.

- I truly enjoyed working with Andreas Henig, Daniel Jung, and Rainer Hörlein, whom I spent with many months of laser beam time adventures abroad, including many unforgettable road trips.
- I am very grateful to Brendan Dromey, Sven Steinke, and Cort Gautier who have been great collaborators and who helped me going through stressful weeks of laser beam time.
- Many thanks to my coworkers Jianhui Bin, Klaus Allinger, Wenjun Ma, and Peter Hilz for their unlimited support and fun discussions on any subject. I also want to thank Johannes Wenz and Konstantin Khrennikov, who were willing to help me out many times in the lab. I would like to thank all other colleagues of the high field group, it has been a lot of fun and a great experience to work with you.
- I would like to thank the LMU Workshop and Johannes Wulz, who went with me through the plasma mirror project and who have been patient with me and my never-ending, but always changing tasks. I also want to thank Jerzy Szerypo, Hans-Jörg Maier, and Dagmar Frischke for their support in target preparation and Reinhardt Satzkowski for driving me or my equipment wherever I wanted.
- I acknowledge the International Max-Planck Research School on Advanced Photon Science for great support and M. Wild for organizing many enjoyable meetings.
- My special thanks to my friend Alexander Buck for helping me out countless times, and who has been a reliable companion over my whole physics career starting from the first day as an undergraduate student.
- Last, but not the least, I would like to thank my parents for their love and endless support over all these years.

Contents

Chapter 1
Introduction

Soon after the first demonstration of the laser [1], the quest for a coherent light source at even shorter wavelengths emerged. Nowadays, intense, brilliant X-ray beams are obtained from large-scale synchrotrons and have become an indispensable tool in many areas of science and technology. These intense X-ray light sources allow resolving matter on the atomic level, give novel opportunities to condensed matter physics, enable the analysis of large biomolecules and thus help developing new materials or future drugs. Recently, free electron lasers have started operating in the X-ray regime providing X-ray pulses of unprecedented high brightness exceeding those from conventional synchrotron sources by orders of magnitude and now offering time resolution on the femtosecond scale [2, 3]. These next generation light sources are now being built at several laboratories around the globe and will open a new era in many fields of science. However, due to their large cost and size, the number of those facilities will be naturally limited to only a few.

The generation of intense (or even laser-like) XUV or X-ray radiation on a much smaller scale has challenged researchers over decades. A promising route is the scattering of a visible laser pulse from a relativistic electron beam. This scheme relies on the relativistic Doppler effect, which causes a frequency shift in the backscattered photon signal by a factor of $4\gamma^2$, where $\gamma = (1 - \beta^2)^{-1/2}$ is the Lorentz factor of the electron beam. Thus, the radiation produced can in principle be tuned freely by varying the energy of the electron beam. Compared to synchrotron or undulator radiation, electrons of rather low kinetic energies are required, which allows reducing the size of the facility considerably.

The concept of using a high energetic electron beam from an accelerator to frequency upshift photons from the visible to the XUV or X-ray range was proposed more than half a century ago [4–6] and has been envisioned as a promising route towards producing intense short wavelength radiation every since. Over the last two decades, this scheme has advanced considerably triggered by major developments in laser and accelerator technology. For instance, high quality, 30 keV X-ray beams were demonstrated via the scattering of a terawatt laser pulse from a conventional electron beam [7]. More recently, all optical configurations using electronbeams

© Springer International Publishing Switzerland 2015
D. Kiefer, *Relativistic Electron Mirrors*, Springer Theses,
DOI 10.1007/978-3-319-07752-9_1

generated from laser plasma accelerators have become subject to experimental [8] and theoretical investigations [9]. Moreover, newly developed compact electron storage rings were combined with high power optical enhancement cavities and are now commercially available as a compact, tunable bright X-ray source [10]. Unprecedented bright γ-ray beams ($\sim$1 MeV) based on Compton backscattering with Doppler upshift $>10^6$ are now being developed (MEGa-ray [11, 12]) and will serve in future as the γ-ray source for the ELI Nuclear Photonics project. These recent achievements are very promising with regard to the development of an intense, tunable X-ray source that fits to the university-laboratory scale. However, owing to the long bunch duration from conventional (or even laser plasma) accelerators, the radiation obtained from these sources is incoherent.

On the contrary, attosecond short, coherent radiation with orders of magnitude higher brightness could be achieved from the coherent backscattering from an extremely short, dense electron bunch, if the thickness of the bunch is small compared to the wavelength of the backscattered radiation. The radiation properties obtained from the coherent backscattering, i.e. the mirror-like reflection, from such electron bunches are intriguing and were first formulated by Einstein, who discussed the reflection from a relativistic mirror as a working example in his paper on special relativity [13]. Upon reflection, the frequency and the amplitude of the incident electromagnetic wave are enhanced by $4\gamma^2$, whereas the pulse is compressed in time by $1/4\gamma^2$, overall resulting in a drastic increase in the peak brightness of the back-reflected electromagnetic pulse. The light pulses that could be generated from the reflection off a relativistic mirror are impressive. For example, if a mirror with $\gamma = 10$ could be produced a laser pulse with a duration of 10 fs and a wavelength of 800 nm would be upshifted to a wavelength of 2 nm and compressed to a pulse duration of 25 as.

While theoretically extremely rewarding, the generation of a relativistic structure that could act as a mirror is very demanding. The advent of high intensity lasers allows the generation of coherent, relativistic structures on a micro-scale that can act as a mirror. Various different schemes have been developed to create a relativistic mirror structure from the interaction of a high intensity laser with a gaseous or solid density plasma. Most prominent example is the high harmonic generation from a relativistically oscillating mirror due to its ability to generate bright attosecond pulses. However, in this case, the mirror acts in an oscillatory mode and hence the generated radiation intrinsically is very broadband [14–16]. Achieving a controlled narrowband upshift requires the generation of a mirror structure propagating with constant velocity. A technique that was successfully demonstrated in experiment is to the reflect off a density spike formed in a laser-driven plasma wakefield, generated in an underdense plasma [17]. In these studies, the reflection from a plasma density wave with $\gamma \sim 5$ was deduced from the observed backscattered signal [18, 19]. However, this approach is limited by the fact that the gamma factor of the density spike in the plasma is determined by the group velocity of the laser in the medium. Thus, reaching high gamma factors requires decreasing the density of the plasma, which however reduces the reflectivity of the mirror structure.

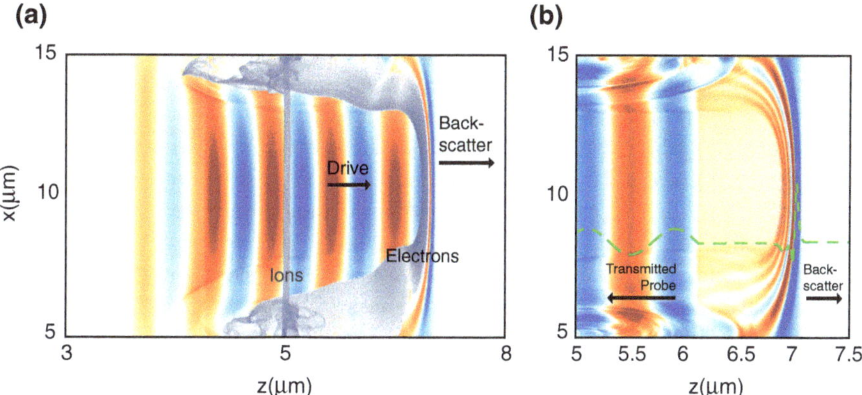

Fig. 1.1 *Laser-driven, relativistic electron mirror (REM) from a nanoscale foil* **a** An idealized high intensity laser pulse, which rises to the peak intensity (5×10^{21} W/cm^2) over one single optical cycle, drives out all electrons from the nanofoil in a single, dense, relativistic electron bunch, which a counter-propagating probe pulse (5×10^{15} W/cm^2) reflects from. **b** Electric field component of the probe pulse (zoom in). The backscattered pulse is frequency upshifted, enhanced in its amplitude and temporally compressed (Pulse propagation directions: Drive: *left* to *right*, Probe: *right* to *left*)

On the contrary, the interaction of a high intensity laser pulse with a nanoscale foil has raised great interest as in this scheme, a freely propagating relativistic structure with remarkably high density could be generated. In the limit of extremely fast rising pulses, it was shown in simulation that all electrons within the nanometer foil could be blown out, at once, in a single, coherent electron bunch, which fully separates from the ions and co-propagates with the accelerating laser field over long distances in vacuum [20]. Numerical studies [21, 22] suggest that attosecond short, relativistic electron layers with density close to solid could be achieved, which truly act as a relativistic mirror and frequency shift counter-propagating light coherently (Fig. 1.1). However, these theoretical studies are highly idealized using step-like rising laser pulses and intensities beyond those available today. In contrast, the formation of dense electron bunches in more realistic interaction scenarios using existing laser technology is largely unexplored and will be investigated in the framework of this thesis.

1.1 Thesis Outline

The aim of this thesis is to investigate the relativistic electron dynamics in high intensity laser–nanofoil interactions. Particular interest is given to the prospect of generating an extremely dense electron bunch that could act as a relativistic mirror and frequency upshift counter-propagating light coherently. This thesis is structures as follows:

Chapter 2 introduces the theoretical framework needed to discuss the experimental results. First, the single electron motion in a strong laser field is derived analytically and the electron dynamics in laser-solid-plasma interactions is discussed. Second, the concept of electron mirror creation from nm scale foils is reviewed, the frequency upshift is derived and the reflection process from laser generated electron mirrors is explained in the framework of coherent scattering theory.

Chapter 3 describes the experimental methods. A short introduction to high power laser systems is given and the key characteristics of laser pulse contrast are discussed. The nanometer thin foils used for the experimental studies are described as well as the diagnostics developed to study the interactions.

Chapter 4 summarizes the electron measurements obtained from different laser systems and target thicknesses. Different interaction regimes are found and explained with the aid of PIC simulations. Experimental data demonstrating the generation of quasi-monoenergetic electron beams from laser–nanofoil interactions is presented and theoretically discussed.

Chapter 5 reports on the dual beam experiment investigating the coherent backscattering from laser-driven electron mirrors. This chapter describes the experimental setup and presents the observed backscatter signal from different interaction configurations. The experimental findings are compared to PIC simulations and an in-depth analysis of the reflection process is given.

Chapter 6 summarizes the results and discusses future perspectives.

Appendix A plasma mirrors for laser pulse contrast enhancement are discussed and different experimental configurations are described. The ATLAS Plasma Mirror design is presented in great detail.

Appendix B supplementary information on the employed spectrometer setups is given and a newly designed wide angle electron ion spectrometer is described.

References

1. Maiman TH (1960) Stimulated optical radiation in ruby. Nature 187(4736):493–494
2. Emma P, Akre R, Arthur J, Bionta R, Bostedt C, Bozek J, Brachmann A, Bucksbaum P, Coffee R, Decker FJ, Ding Y, Dowell D, Edstrom S, Fisher A, Frisch J, Gilevich S, Hastings J, Hays G, Hering Ph, Huang Z, Iverson R, Loos H, Messerschmidt M, Miahnahri A, Moeller S, Nuhn HD, Pile G, Ratner D, Rzepiela J, Schultz D, Smith T, Stefan P, Tompkins H, Turner J, Welch J, White W, Wu J, Yocky G, Galayda J (2010) First lasing and operation of an angstrom-wavelength free-electron laser. Nat Photonics 4(9):641–647
3. W. Ackermann and Others (2007) Operation of a free-electron laser from the extreme ultraviolet to the water window. Nat Photonics, 1(6):336–342
4. Landecker K (1952) Possibility of frequency multiplication and wave amplification by means of some relativistic effects. Phys Rev 86:852–855
5. Milburn Richard H (1963) Electron scattering by an intense polarized photon field. Phys Rev Lett 10:75–77
6. Arutyunian FR (1963) The compton effect on relativistic electrons and the possibility of obtaining high energy beams. Phys Lett 4:176–178
7. Schoenlein RW, Leemans WP, Chin AH, Volfbeyn P, Glover TE, Balling P, Zolotorev M, Kim KJ, Chattopadhyay S, Shank CV (1996) Femtosecond x-ray pulses at 0.4 A generated by 90 ° thomson scattering: a tool for probing the structural dynamics of materials. Science 274(5285):236–238

8. Schwoerer H, Liesfeld B, Schlenvoigt H-P, Amthor K-U, Sauerbrey R (2006) Thomson-backscattered x rays from laser-accelerated electrons. Phys Rev Lett 96(1):014802
9. Hartemann FV, Gibson DJ, Brown WJ, Rousse A, Ta Phuoc K, Mallka V, Faure J, Pukhov A (2007). Compton scattering x-ray sources driven by laser wakefield acceleration. Phys Rev ST Accel Beams 10:011301
10. Zhirong Huang, Ruth Ronald D (1998) Laser-electron storage ring. Phys Rev Lett 80:976–979
11. Brown WJ, Anderson SG, Barty CPJ, Betts SM, Booth R, Crane JK, Cross RR, Fittinghoff DN, Gibson DJ, Hartemann FV, Hartouni EP, Kuba J, Le Sage GP, Slaughter DR, Tremaine AM, Wootton AJ, Springer PT, Rosenzweig JB (2004) Experimental characterization of an ultrafast thomson scattering x-ray source with three-dimensional time and frequency-domain analysis. Phys Rev ST Accel Beams 7:060702
12. Albert F, Anderson SG, Anderson GA, Betts SM, Gibson DJ, Hagmann CA, Hall J, Johnson MS, Messerly MJ, Semenov VA, Shverdin MY, Tremaine AM, Hartemann FV, Siders CW, McNabb DP, Barty CPJ (2010) Isotope-specific detection of low-density materials with laser-based monoenergetic gamma-rays. Opt Lett 35(3):354–356
13. Einstein A (2005) Zur Elektrodynamik bewegter Körper [AdP 17, 891 (1905)]. Ann Phys 14(S1):194–224
14. Dromey B, Zepf M, Gopal A, Lancaster K, Wei MS, Krushelnick K, Tatarakis M, Vakakis N, Moustaizis S, Kodama R, Tampo M, Stoeckl C, Clarke R, Habara H, Neely D, Karsch S, Norreys P (2006) High harmonic generation in the relativistic limit. Nat Phys 2(7):456–459
15. Dromey B, Kar S, Bellei C, Carroll DC, Clarke RJ, Green JS, Kneip S, Markey K, Nagel SR, Simpson PT, Willingale L, McKenna P, Neely D, Najmudin Z, Krushelnick K, Norreys PA, Zepf M (2007) Bright multi-kev harmonic generation from relativistically oscillating plasma surfaces. Phys Rev Lett 99(8):085001
16. Baeva T, Gordienko S, Pukhov A (2006) Theory of high-order harmonic generation in relativistic laser interaction with overdense plasma. Phys Rev E 74:046404
17. Bulanov Sergei V, Timur Esirkepov, Toshiki Tajima (2003) Light intensification towards the schwinger limit. Phys Rev Lett 91(8):085001
18. Kando M, Fukuda Y, Pirozhkov AS, Ma J, Daito I, Chen L-M, Esirkepov TZh, Ogura K, Homma T, Hayashi Y, Kotaki H, Sagisaka A, Mori M, Koga JK, Daido H, Bulanov SV, Kimura T, Kato Y, Tajima T (2007) Demonstration of laser-frequency upshift by electron-density modulations in a plasma wakefield. Phys Rev Lett 99(13):135001
19. Kando M, Pirozhkov AS, Kawase K, Esirkepov TZh, Fukuda Y, Kiriyama H, Okada H, Daito I, Kameshima T, Hayashi Y, Kotaki H, Mori M, Koga JK, Daido H, Faenov AY, Pikuz T, Ma J, Chen LM, Ragozin EN, Kawachi T, Kato Y, Tajima T, Bulanov SV (2009) Enhancement of photon number reflected by the relativistic flying mirror. Phys Rev Lett 103(23):235003
20. Kulagin VV, Cherepenin VA, Hur MS, Suk H (2007). Theoretical investigation of controlled generation of a dense attosecond relativistic electron bunch from the interaction of an ultrashort laser pulse with a nanofilm. Phys Rev Lett 99(12):124801
21. Meyer-ter Vehn J, Wu HC (2009) Coherent thomson backscattering from laser-driven relativistic ultra-thin electron layers. Eur Phys J D 55:433–441
22. Qiao B, Zepf M, Borghesi M, Dromey B, Geissler M (2009) Coherent x-ray production via pulse reflection from laser-driven dense electron sheets. New J Phys 11(10):103042 (11pp)

Chapter 2
Theoretical Background

A high intensity laser pulse ($>10^{18}\ \mathrm{W/cm^2}$) incident on a nanometer thin foil rapidly ionizes the atoms of the irradiated material and thus interacts with a solid density plasma. The ionization process sets in at comparably low intensities ($\sim 10^{13}\ \mathrm{W/cm^2}$) at the foot of the pulse, and in strong fields, is well described through tunnel or barrier suppression ionization, covered by many textbooks [1]. This chapter introduces the theoretical framework needed to understand the electron dynamics in laser plasma interactions, reviews the concept of electron mirror generation from nanoscale foils and discusses the reflection properties of relativistic electron mirror structures.

2.1 Fundamentals of Light

Electromagnetic radiation is described by Maxwell's equations [2]. The electric and magnetic fields $\boldsymbol{E}$, $\boldsymbol{B}$ can be directly found from them. Introducing the potentials $\boldsymbol{A}$, ϕ such that

$$\begin{aligned} \boldsymbol{E} &= -\nabla\phi - \tfrac{\partial}{\partial t}\boldsymbol{A} \\ \boldsymbol{B} &= \nabla \times \boldsymbol{A} \end{aligned} \tag{2.1}$$

and using the Lorenz Gauge $\nabla \boldsymbol{A} + c^{-2}\partial\phi/\partial t = 0$, Maxwell's equations reduce to the symmetric wave equations

$$\begin{aligned} \Delta\phi - \tfrac{1}{c^2}\tfrac{\partial^2}{\partial t^2}\phi &= -\rho/\epsilon_0 \\ \Delta\boldsymbol{A} - \tfrac{1}{c^2}\tfrac{\partial^2}{\partial t^2}\boldsymbol{A} &= -\mu_0\boldsymbol{j} \end{aligned} \tag{2.2}$$

where c denotes the speed of light, ϵ_0 the electric permittivity and μ_0 magnetic permeability. In vacuum, the electric charge and current density vanish ($\boldsymbol{j} = \rho = 0$) and hence, a laser pulse is simply described by

$$\boldsymbol{A}(\boldsymbol{r}, t) = \boldsymbol{A}_A(\boldsymbol{r}, t)\sin(\boldsymbol{k}_L \cdot \boldsymbol{r} - \omega_L t + \phi) \tag{2.3}$$

© Springer International Publishing Switzerland 2015

D. Kiefer, *Relativistic Electron Mirrors*, Springer Theses,
DOI 10.1007/978-3-319-07752-9_2

with the dispersion relation $\omega_L = ck_L$ and phase ϕ. Thus, the electric and magnetic fields are given by

$$\begin{aligned} \boldsymbol{E}(\boldsymbol{r},t) &= \boldsymbol{E_A}(\boldsymbol{r},t)\cos(\boldsymbol{k_L}\cdot\boldsymbol{r}-\omega_L t+\phi) \\ \boldsymbol{B}(\boldsymbol{r},t) &= \boldsymbol{B_A}(\boldsymbol{r},t)\cos(\boldsymbol{k_L}\cdot\boldsymbol{r}-\omega_L t+\phi) \end{aligned} \tag{2.4}$$

with envelope functions $\boldsymbol{E_A} = c\boldsymbol{B_A} = \omega_L \boldsymbol{A_A}$ and $\boldsymbol{E_A} \perp \boldsymbol{B_A}$, $\boldsymbol{E_A} \perp \boldsymbol{k_L}$, $\boldsymbol{B_A} \perp \boldsymbol{k_L}$. For a plane wave, $E_A(\boldsymbol{r},t) = E_0$, whereas for a gaussian pulse shape, the field distribution in the focal point is $E_A(\boldsymbol{r},t) = E_0\, e^{-t^2/\tau_L^2}\, e^{-(x^2+y^2)/w_0^2}$. Assuming a gaussian profile (in space and time), the peak intensity of the pulse can be determined from the laser pulse energy E, the FWHM pulse duration t_{FWHM} and the FWHM focal spot size d_{FWHM} using[1]

$$I_0 = \frac{0.82\cdot E}{t_{FWHM}\, d_{FWHM}^2} \tag{2.5}$$

Theoretically, the intensity of the pulse can be derived from the cycle-averaged Poynting vector, thus $I_0 = \langle S\rangle_T = \epsilon_0 c^2 \langle |E\times B|\rangle_T = c\epsilon_0 E_0^2/2$. Now, if we use the normalized vector potential $\boldsymbol{a} = e\boldsymbol{A}/m_e c$ to express the electric field of the laser $E_0 = m_e c\omega_L/e\cdot a_0$ we find for the intensity

$$I_0 = \frac{1.37\cdot 10^{18}\,\mathrm{W/cm^2}}{\lambda^2[\mu\mathrm{m}]} a_0^2 \tag{2.6}$$

Using that expression in combination with Eq. 2.5, we can deduce the a_0 parameter frequently used in theory and simulation. It is worth noting that the fields achieved with the laser pulse are simply

$$\begin{aligned} E_L &= 3.2\cdot \tfrac{a_0}{\lambda_L[\mu\mathrm{m}]} \times 10^{12}\,\mathrm{V/m} \\ B_L &= 1.07\cdot \tfrac{a_0}{\lambda_L[\mu\mathrm{m}]} \times 10^{4}\,\mathrm{T} \end{aligned} \tag{2.7}$$

Thus, the laser pulses used in this thesis reach electric fields in the range of tens of TV/m and magnetic fields on the order of 10^4–10^5 T.

2.2 Single Electron Motion in a Relativistic Laser Field

The interaction of an intense laser pulse with a solid density plasma is a very complex, many body system, which in general cannot be described analytically. Nonetheless, to get a better insight into the interaction dynamics, it is instructive to study the single electron motion in an electromagnetic wave, as these dynamics very often can still be recovered even in the large scale systems.

[1] $t_{FWHM} = \sqrt{2\ln 2}\tau_L$, $d_{FWHM} = \sqrt{2\ln 2}w_0$.

The equation of motion of an electron in an electromagnetic field is given by the Newton-Lorentz equation

$$\frac{d}{dt}\boldsymbol{p} = -e\,(\boldsymbol{E} + \boldsymbol{v} \times \boldsymbol{B}) \tag{2.8}$$

This set of coupled partial differential equations can by solved analytically following [3, 4]. However, a deeper understanding of the system can be gained using the Lagrangian formalism and considering fundamental symmetries [5].

2.2.1 Symmetries and Invariants

In the following, we will work in relativistic units. The normalized variables are derived from their counterparts in SI-units:

$$E \rightarrow E' = \frac{E}{m_e c^2} \qquad \Phi \rightarrow \phi' = \frac{e\,\Phi}{m_e c^2} \qquad z \rightarrow z' = k_L z$$
$$p \rightarrow p' = \frac{p}{m_e c} \qquad A \rightarrow a' = \frac{e\,A}{m_e c} \qquad t \rightarrow t' = \omega_L t$$

Note that the energy of the particle is just $E = \gamma$ with $\gamma = (1 - \boldsymbol{\beta}^2)^{-1/2} = \sqrt{1 + {p'_x}^2 + {p'_z}^2}$. For the sake of simplicity we shall neglect the $'$ in the following discussion. The relativistic Lagrangian function of an electron moving in an electromagnetic field with vector potential $\boldsymbol{A}$ and electrostatic potential ϕ reads [2, 6]

$$L = -\sqrt{1 - \boldsymbol{\beta}^2} - \boldsymbol{\beta}\boldsymbol{a} + \phi \tag{2.9}$$

from which we can derive the canonical momentum $\boldsymbol{p}^{can} = \frac{\partial L}{\partial \boldsymbol{\beta}} = \gamma\boldsymbol{\beta} - \boldsymbol{a} = \boldsymbol{p} - \boldsymbol{a}$. If we now consider potentials that are dependent on the z coordinate only, i.e. $\boldsymbol{a} = a(z,t)\,\boldsymbol{e_x}$ and $\phi = \phi(z)$, the planar symmetry of the system $\partial L/\partial x = 0$ implies that the canonical momentum in the transverse direction is conserved, that is

$$\frac{d}{dt}p_x^{can} = \frac{d}{dt}\frac{\partial L}{\partial \beta_x} = \frac{\partial L}{\partial x} = 0 \Rightarrow p_x - a = const \tag{2.10}$$

We can derive a second invariant if we neglect the electrostatic potential $\phi = 0$ and consider a wave form $a = a(t - z)$. As a result, the system is anti-symmetric in the coordinates z, t, which implies $\partial L/\partial t = -\partial L/\partial z$. Making use of the relation $dH/dt = -\partial L/\partial t$ for the Hamiltonian function, we can write

$$\frac{dH}{dt} = -\frac{\partial}{\partial t}L = \frac{\partial L}{\partial z} = \frac{d}{dt}\frac{\partial L}{\partial \beta_z} = \frac{d}{dt}p_z^{can} \tag{2.11}$$

and we find the second integral of motion ($H = \gamma$)

$$\frac{d}{dt}\left(\gamma - p_z^{can}\right) = 0 \Rightarrow \gamma - p_z^{can} = const. \tag{2.12}$$

We can now immediately solve the equations of motion making use of the integrals derived in the previous section. Conservation of the transverse canonical momentum (Eq. 2.10) yields $p_x^{can}(t) = p_x^{can}(t_0) = \alpha_0$, hence

$$p_x(t) = \alpha_0 + a(t) \tag{2.13}$$

As for a plane wave $a_z = 0$, thus $p_z^{can} = p_z$, we define the constant of motion $\kappa_0 = (\gamma - p_z)\,|_{t=t_0}$ and obtain from the second invariant (Eq. 2.12)

$$\gamma(t) = \kappa_0 + p_z(t) \tag{2.14}$$

which in combination with $\gamma = \sqrt{1 + p_x^2 + p_z^2}$ gives

$$p_z(t) = \frac{1}{2\kappa_0}\left(1 - \kappa_0^2 + p_x^2(t)\right) \tag{2.15}$$

Now, if we consider a plane wave with electric field $e_L = -a_0 \cos(\tau + \phi_0)$ and vector potential $a = a_0 \sin(\tau + \phi_0)$, where $\tau = t - z$, we immediately find for the momenta

$$\begin{aligned} p_x(\tau) &= \gamma\beta_\perp = a_0 \sin(\tau + \phi_0) + \alpha_0 \\ p_z(\tau) &= \gamma\beta_z = \tfrac{1}{2\kappa_0}\left(1 - \kappa_0^2 + [a_0 \sin(\tau + \phi_0) + \alpha_0]^2\right) \\ \gamma(\tau) &= \kappa_0 + \tfrac{1}{2\kappa_0}\left(1 - \kappa_0^2 + [a_0 \sin(\tau + \phi_0) + \alpha_0]^2\right) \end{aligned} \tag{2.16}$$

where the constants of motion α_0, κ_0 can be determined from the initial conditions $p_{z,0}$, $p_{x,0}$, ϕ_0

$$\alpha_0 = p_{x,0} - a_0 \sin\phi_0 \qquad \kappa_0 = \gamma_0 - p_{z,0} \qquad \gamma_0 = \sqrt{1 + p_{\perp,0}^2 + p_{z,0}^2} \tag{2.17}$$

To obtain the electron trajectory, we make use of a change in variables which considerably simplifies the integration of Eq. 2.16. Using $\tau = t - z$ as independent variable implies $d\tau = (1 - \beta_z)\,dt = \kappa_0/\gamma\,dt$,[2] thus substitution gives $dz/d\tau = \gamma/\kappa_0\,dz/dt = p_z/\kappa_0$ and $dx/d\tau = \gamma/\kappa_0\,dx/dt = p_x/\kappa_0$, which can be integrated

[2] $(1 - \beta_z) = (\gamma - p_z)/\gamma = \kappa_0/\gamma$, using Eq. (2.14).

$$t = z + \tau$$
$$x(\tau) = \frac{1}{\kappa_0}\left[\alpha_0 \tau - a_0 \left(\cos\left(\tau + \phi_0\right) - \cos\phi_0\right)\right] \tag{2.18}$$
$$z(\tau) = \frac{1}{\kappa_0^2}\left[\left(1 + \alpha_0^2 - \kappa_0^2 + \frac{a_0^2}{2}\right)\frac{\tau}{2} - a_0\left(\alpha_0 \cos\left(\tau + \phi_0\right) + \frac{a_0}{8}\sin\left(2\left(\tau + \phi_0\right)\right)\right)\right.$$
$$\left. + a_0\left(\alpha_0 \cos\phi_0 + \frac{a_0}{8}\sin 2\phi_0\right)\right]$$

It is worth noting that for an electron initially at rest ($p_{z,0} = p_{x,0} = 0$), Eqs. 2.13–2.15 simplify considerably, as in this case

$$\begin{aligned} p_x(t) &= a(t) - a(t_0) \\ p_z(t) &= \tfrac{1}{2}p_x^2(t) \\ \gamma(t) &= 1 + p_z(t) \end{aligned} \tag{2.19}$$

Hence, the kinetic energy is just $E_{kin} = (\gamma - 1) = p_x^2/2$, which reveals that the energy gain of the particle stems from the transverse electric field, whereas the $v \times B$ term turns the particle quiver motion into the forward direction without adding energy to it.

Figure 2.1 depicts the electron dynamics of an electron initially at rest. The particle motion is strongly dependent on the initial phase, which crucially governs the maximal energy achieved in the field $\gamma^{max} = 1 + \frac{a_0^2}{2}\left(1 + \sin\phi_0\right)^2$. Moreover, depending on the initial phase, the electron oscillates in transverse dimension with amplitude $x^{max} = a_0$ or gradually drifts in either one direction (Fig. 2.1d).

2.2.2 Single Electron Motion in a Finite Pulse

The solution derived so far is strictly speaking only valid for infinite plane waves. Imposing a more realistic temporally finite, gaussian shaped pulse the equations of motion cannot be solved analytically anymore and numerical methods (here: Fourth Order Runge-Kutta) need to be used. Figure 2.2 shows the numerical integration of an electron propagating in a gaussian shaped, finite pulse. The kinetic energy of the electron is directly coupled to the light field and returns back to zero as soon as the (slightly slower propagating) electron is overtaken by the laser pulse. This is a direct consequence of the conservation of the transverse canonical momentum (Eq. 2.10). Since initially $p_x(t = -\infty) = a(t = -\infty) = 0$ the final transverse momentum is $p_x(t = \infty) = a(t = \infty) = 0$ and likewise $p_z = p_x^2/2 = 0$, which means that a charged particle cannot gain energy from a plane wave in vacuum. It needs the break up of symmetry for effective energy gain.

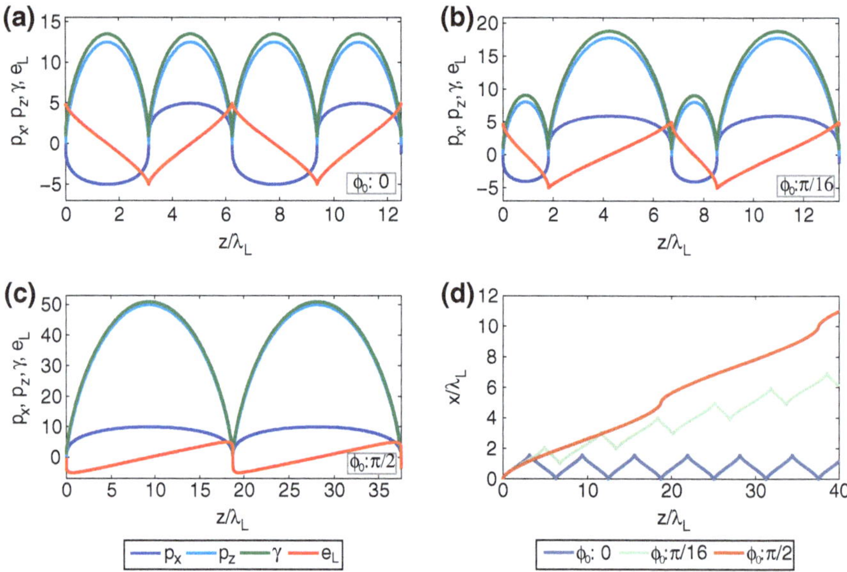

Fig. 2.1 *Single electron motion in a plane wave*. Depending on the injection phase ϕ_0, the electron is accelerated (decelerated) within one quarter ($\phi_0 = 0$) to one half cycle ($\phi_0 = \pi/2$) of the driving field ($a_0 = 5$). Note the different scales of the abscissa and ordinate axis. **a**–**c** depict the electron slippage over 2 laser cycles, **d** shows the corresponding electron motion in space

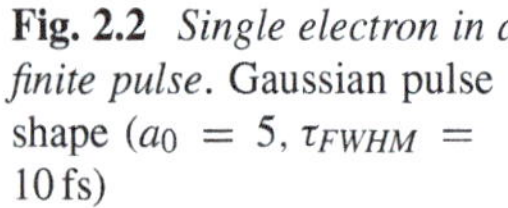

Fig. 2.2 *Single electron in a finite pulse*. Gaussian pulse shape ($a_0 = 5$, $\tau_{FWHM} = 10\,\text{fs}$)

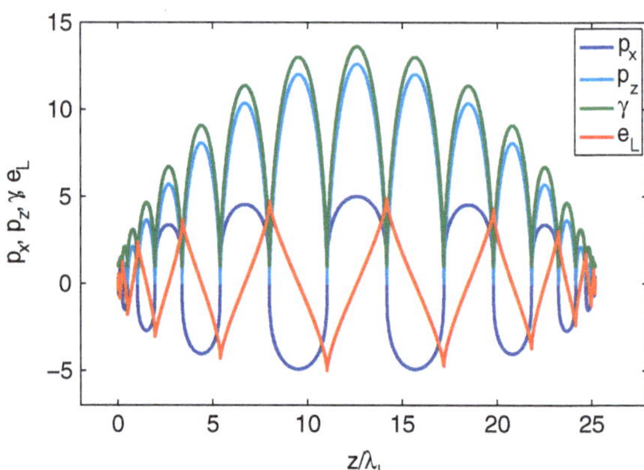

2.2.3 The Lawson Woodward Principle and Its Limitations

The fundamental question under what conditions a free electron can extract energy from an electromagnetic laser field has been a controversial debate over many years. General starting point is the so called Lawson-Woodward Theorem [7, 8], which states that the net energy gain of an isolated relativistic electron interacting with an electromagnetic field is zero. However, the proof of this theorem is bound to a number of assumptions [1, 9, 10]

- the laser field is in vacuum with no walls or boundaries present,
- the electron is ultra-relativistic along the acceleration path,
- no static electric or magnetic fields are present,
- the interaction region is infinite,
- nonlinear effects can be neglected.

Here, it should be noted that the Lorentz force $\boldsymbol{v} \times \boldsymbol{B}$ is linear in the ultra-relativistic case ($v \rightarrow c$) and does not violate the Lawson-Woodward Principle. Despite the vast number of underlying assumptions, this theorem has proven its relevance over the years and was recently confirmed in a test experiment [11].

Nonetheless, numerous acceleration schemes have been developed in theory violating one or many of the underlying conditions in order to accelerate electrons in vacuum effectively. In the following, we will highlight only a few aspects of those schemes relevant for this work.

2.2.4 Acceleration in an Asymmetric Pulse

Breaking up symmetry in time and assuming that we could find a mechanism that could inject electrons right into the middle of a pulse at time t_0, the situation completely changes and a non-zero energy gain can be extracted from the electromagnetic field [4, 12]. Using Eq. 2.19, we find for the final energy of the electron

$$\gamma^{final} = 1 + \frac{1}{2}(a(\infty) - a(t_0))^2 = 1 + \frac{1}{2}a(t_0)^2 \tag{2.20}$$

Thus, the energy gain strongly depends on the phase of the field at the injection time t_0. Approximating the vector potential of a gaussian pulse with $a \approx a_0 \exp(-t^2/\tau_L^2)\sin\phi(t, x)$ (adiabatic approximation) and taking into account that the electric field is $e_L = -\partial a/\partial t$, we find maximum energy gain for $\phi(t, x) = \pi/2$ corresponding to $e_L \propto \cos(\pi/2) = 0$. Hence, electrons injected into the field at the zero points close to the peak of the pulse experience substantial energy gain from the electromagnetic field as can be seen in Fig. 2.3.

A scheme that could potentially seed electrons right into the peak of the pulse is to exploit the ionization dynamics of highly charged ions [13, 14]. As it was shown in simulation, inner shell electrons of high Z atoms remain during the rise time of the laser pulse and are released from the ionic core (and thus injected right into the maximal intensity region) when the pulse reaches its peak intensity. Recently, it was pointed out that the laser nanofoil interaction might exhibit similar dynamics, which could provide effective means of accelerating electrons from semi-transparent solid plasmas and which will be discussed in great detail in Chap. 4.

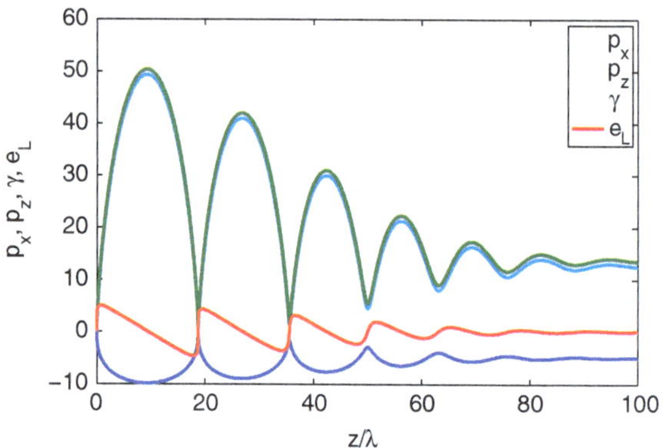

Fig. 2.3 *Single electron in an asymmetric pulse*. The electron is injected into the laser field at the peak of the pulse with $p_{x,0} = p_{z,0} = 0$ and $\phi_0 = \pi/2$. Same pulse as in Fig. 2.2

2.2.5 Ponderomotive Scattering

To reach high intensities, laser pulses are focused tightly to within a few μm only and thus the field distribution interacting with the electron in experiment is strongly dependent on its radial position. While for a plane wave, the cycle-averaged Lorentz force acting on the particle turns out to be zero,[3] inhomogeneous fields exhibit a nonzero component, which causes the particle to drift from high intensity to low intensity regions. The origin of the ponderomotive force can be easily understood if we consider a particle initially located at the center of the focal spot. Owing to the transverse electric field, the electron is displaced from its central position to regions of lowered intensities. Thus, as the oscillating field changes sign the force driving the electron back to the center is smaller and therefore, the electron does not return to its initial position. As a result, the oscillation center gradually drifts from regions of high intensity to those of lower intensity while the mean kinetic energy of the particle successively increases with every cycle.

This phenomenon is well known at sub-relativistic intensities and can be derived from first order perturbation analysis of the Lorentz force around the oscillation center [1].[4] In the relativistic regime, the longitudinal motion has to be taken into account. Assuming that the particle motion can be separated into $\boldsymbol{p} = \bar{\boldsymbol{p}} + \tilde{\boldsymbol{p}}$ where $\bar{\boldsymbol{p}}$ and $\tilde{\boldsymbol{p}}$ denote the slowly varying and the rapidly varying part with respect to the laser frequency, the generalized, relativistic ponderomotive force reads [15, 16]

$$\boldsymbol{F_p} = -\frac{m_e c^2}{4\bar{\gamma}} \nabla \boldsymbol{a_A}^2 \qquad \bar{\gamma} = \sqrt{1 + \bar{p}_z^2 + \bar{p}_\perp^2 + a_A^2/2} \tag{2.21}$$

---

[3] $F \propto p/\gamma \cdot B \propto \sin\tau \cos\tau \propto \sin 2\tau$, thus $\langle F \rangle_\tau = 0$.

[4] At sub-relativistic intensities, the ponderomotive potential of the laser field is $\Phi_p = \frac{e^2 E_A^2}{4 m_e \omega_L^2} = \frac{m_e c^2}{4} a_A^2$ and the ponderomotive force is just simply $F_p = -\nabla \Phi_p = -\frac{m_e c^2}{4} \nabla a_A^2$.

The main feature still applies: The electron drifts away from the high intensity region owing to the gradient of the intensity distribution and eventually scatters out of the focused beam—thus, overall gaining energy from the electromagnetic field of the laser (Fig. 2.4). While this process was observed in experiment at rather low intensities, accelerating electrons up to few hundred keVs and scattering angles in excellent agreement with those expected from single electron dynamics [17, 18], the ponderomotive scattering in the high intensity regime [19, 20], which is expect to occur when the electron quiver amplitude ($x = a_0$) reaches the length scale of the beam waist at the focus has been discussed quite controversial [16, 21]. In particular, it was shown that a rather simple treatment of the electromagnetic field distribution in the focal plane using the paraxial Gaussian beam approximation [19] fails considerably in predicting the final energy gain and angular distribution [16, 22]. Including higher order corrections, especially longitudinal fields, the final energy gain is found to be significantly reduced, the scattering angle turns out to be highly dependent on the initial position and is no longer limited to the polarization plane only. Taking into account that the actual focal distribution of test particle studies is rather difficult.

Figure 2.4 illustrates the ponderomotive scattering of an electron in a Gaussian mode (lowest order approximation) clearly showing the effective energy gain of an electron from a finite field distribution in space. Longitudinal field components appear in the next order [16] which may play an important role. A correct field distribution up to all orders is given in [16, 22], nonetheless, this may still be different from the actual experimental conditions.

In conclusion, we find that the dynamics of a single electron injected into a relativistic, tightly focused laser pulse is very complex with strong dependence on the exact field distribution in the focal region and the initial position of the electron.

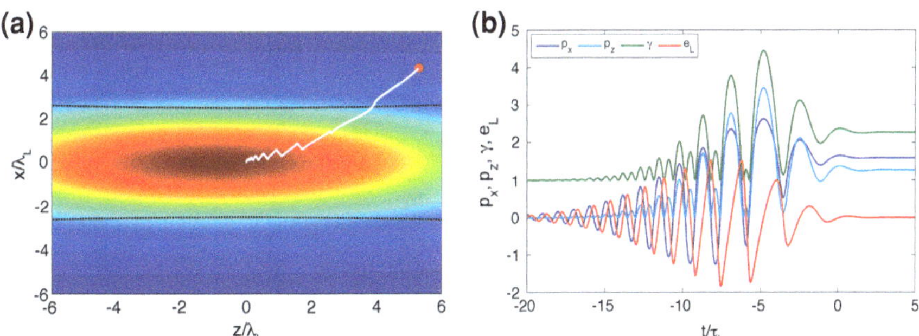

Fig. 2.4 *Ponderomotive scattering*. Single electron in a finite, Gaussian shaped pulse with beam waist $w_0 = 2\,\mu\text{m}$ and pulse duration $\tau_{FWHM} = 10\,\text{fs}$. **a** Electron trajectory (*white line*) and instantaneous position (*red dot*) at $t/\tau_L = -1.1$ superimposed with a snapshot of the cycle-averaged intensity distribution at that moment in time. **b** Temporal evolution of the electron energy and momentum

2.2.6 *Vacuum Acceleration Schemes*

While in the case of ponderomotive scattering, electrons are quickly expelled from the focused laser beam, certain regions surrounding the laser axis have been identified where high energetic electrons can be trapped and accelerated for a long time [23–25]. Detailed analysis of the diffracting laser beam reveals that in these sectors the effective phase velocity of the laser field is slightly smaller than the speed of light. Hence, relativistic electrons injected into these regions are quasi-phase-matched with the accelerating field and thus experience a drastic energy gain. Although it was argued that the so-called electron capture and acceleration scenario (CAS) even works for electrons initially at rest when accounting for the longitudinal field components of the focal spot [22], the mechanism requires rather high intensities $a_0 \sim 10$–100, is critically dependent on the exact field distribution and thus still remains experimentally unexplored.

While in the high intensity regime, electrons initially at rest interacting with a tightly focused beam tend to be scattered transversally long before the peak of the pulse has reached, it was argued that a ring-like intensity profile would focus the accelerating particles towards the beam axis, owing to the off-axis potential well originating from the intensity distribution [26, 27].

2.3 Laser Propagation in a Plasma

We now turn our discussion from single particle interactions to a dense plasma. Here, we shall briefly introduce the fundamental properties of a cold plasma, meaning that we essentially neglect forces arising from the thermal pressure of the plasma. Derivations are given in many textbooks [1, 5, 28].

In a neutral plasma, electrons displaced from their equilibrium position feel a restoring force caused by the positive ion background and thus oscillate with the plasma frequency

$$\omega_p = \sqrt{\frac{n_e e^2}{\epsilon_0 m_e \bar{\gamma}}} \tag{2.22}$$

where $\bar{\gamma}$ is the cycle-averaged Lorentz factor in the plasma, often set to $\bar{\gamma} = \sqrt{1 + a_0^2/2}$. It is worth noting that due to their much higher mass, ions stay quasi immobile on the time scale of the plasma frequency and thus can be viewed as a uniform background in this context. From the dispersion relation of an electromagnetic wave propagating in a plasma,

$$\omega_L^2 = \omega_p^2 + c^2 k_L^2 \tag{2.23}$$

we can derive the refractive index $n_R = c/v_{ph}$

$$n_R = \sqrt{1 - \frac{\omega_p^2}{\omega_L^2}} \tag{2.24}$$

Thus, in the case of a rather low density plasma ($\omega_p < \omega_L$), light propagates with phase velocity $v_{ph} = c/n_R$ and group velocity $v_g = cn_R$. However, if $\omega_p > \omega_L$, the refractive index becomes imaginary. In this case, the response of the plasma electrons is much faster than the frequency of the electromagnetic wave and therefore the incident wave is effectively shielded at every moment in time in the plasma. Depending on the electron density, the plasma can either be overdense (opaque) or underdense (transparent) to the incident light field. The interaction dynamics are fundamentally different in these two scenarios and we define the critical density at which $\omega_p = \omega_L$, to distinguish those two regimes. Using Eq. 2.22, we find for the critical density

$$n_c = \frac{\epsilon_0 m_e}{e^2} \bar{\gamma}\, \omega_L^2 = \bar{\gamma} \cdot \frac{1.1 \cdot 10^{21}}{\lambda[\mu\mathrm{m}]^2} \mathrm{cm}^{-3} \tag{2.25}$$

Hence, an electromagnetic wave incident on an overdense plasma reflects from the plasma surface where it interacts as an evanescent wave within the skin layer of the plasma. For a step-like boundary, we can define the characteristic length scale over which the electric field drops to $1/e$, i.e. the plasma skin depth as

$$l_s = \frac{c}{\sqrt{\omega_p^2 - \omega_L^2}} \approx \frac{c}{\omega_p} \tag{2.26}$$

2.3.1 Laser Interaction with an Overdense Plasma

A laser pulse normally incident on an overdense plasma is reflected and thus interacts as a standing wave with the critical surface of the plasma. At relativistic intensities, the $v \times B$ component of the resultant electromagnetic field drives the plasma surface in longitudinal direction with

$$F_z = F_0 \left(1 - \cos 2\omega_L t\right) \tag{2.27}$$

which oscillates at twice the frequency of the incident laser field.[5] Note, that the driving force does not change sign and thus on time average pushes the critical surface into the plasma, whereas the oscillating high frequency component eventually leads

[5] this rather general expression is readily derived from the ponderomotive force when including the fast oscillating component [29], or alternatively from a perturbative model [30]. A more detailed theoretical treatment is given in [31] using a one particle plasma model, which has proven good agreement with PIC simulations.

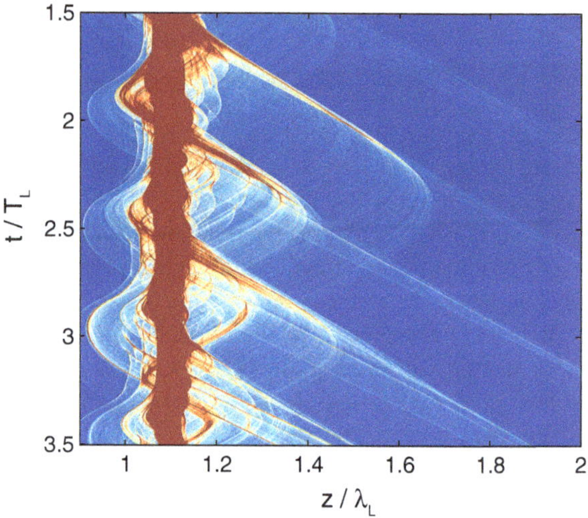

Fig. 2.5 *Electron bunch generation at the laser-plasma boundary*. At every laser (half-)cycle, a group of electrons is accelerated to MeV energies and injected as a dense bunch into the plasma

to strong electron heating. At oblique incidence, the situation is quite similar. Here, the leading term driving the critical surface is the electric field component pointing normal to the plasma boundary, which however oscillates at a frequency of ω_L, only, and acts in both directions. In both cases, the interplay between the driving force and the restoring charge separation field leads to the oscillation of the plasma surface at the frequency of the driving force. This collective motion of the electrons at the plasma boundary can be modeled analytically [31] and is the key component for the generation of high harmonics from solids in the relativistic regime.

Along with the oscillatory surface motion, at every half (full) cycle, a group of electrons acquires high energies at the laser plasma boundary and is injected as a dense bunch into the overdense region (Fig. 2.5). As the laser field does not penetrate into the plasma interior, these electrons immediately escape from the driving laser field with energies on the order of several MeVs well above the bulk electron plasma temperature.

The periodic formation of these high energetic electron bunches at a sharp laser plasma boundary is evident in simulations and has been confirmed experimentally probing the optical transition radiation emitted from the generated hot electron current crossing the rear surface of the target. Here, the optical emission spectra were found to be spiked at ω_L and $2\omega_L$, which hints that these bunches preserve their temporal periodicity to some extend as they propagate through the plasma [32, 33]. In the vacuum region behind the target, the expelled electron bunches quickly disperse in the electrostatic sheath field built up during the interaction and eventually form a hot electron cloud surrounding the target rear side, which in turn causes the acceleration of ions.

Although initially highly confined in space, the generated electron bunches are spectrally very broadband. Moreover, the energy distribution of subsequent bunches fluctuates from cycle to cycle and the overall, time-integrated electron spectra observed in experiment and simulation resemble exponentially decaying distribution functions, with characteristic slope commonly referred to as the hot electron temperature. As it was pointed out by Bezzerides et al. [34], the spectral shape is a direct consequence of the stochastic nature of the bunch formation process, as theoretically, the integration over many bunches with random variations in the energy spectrum eventually leads to a Maxwellian distribution.

While exponential, hot electron distributions have been measured over decades in laser plasma experiments [1, 35–39], the physical mechanism of the electron bunch formation at the vacuum plasma interface is still not understood. Recently, a deeper insight into the process was given by Mulser et al. [40] who showed that this phenomenon may be explained by an anharmonic resonance in the attractive charge separation potential at the plasma vacuum boundary. Here, electrons with large oscillation amplitude may be driven into resonance thereby break up with the collective plasma motion and rapidly gain energy from the laser. Yet, owing to the stochastic nature of this process, no theory exists to date, which could for a given set of parameters make a prediction on the electron number within a bunch or anticipate its energy distribution.

Instead, numerous scalings have been developed predicting the slope of the time-integrated hot electron distribution [41–44]. In the case of a normal incident laser pulse, [41] showed that the hot electron temperature can be related to the ponderomotive energy of the laser pulse

$$k_B T_{hot}^{Wilks} = m_e c^2 \left(\sqrt{1 + a_0^2/2} - 1\right) \tag{2.28}$$

This scaling is intriguingly simple and experimental configurations showing fairly good agreement with the ponderomotive scaling were reported [45]. However, a more recent theoretical study [44] showed that the ponderomotive scaling is actually only valid at sub-relativistic intensities, whereas the scaling increasingly overestimates the hot electron temperatures at intensities clearly beyond the relativistic threshold ($a_0 \gg 1$).Using that the average kinetic energy of an electron ensemble can be obtained by averaging the single electron energy with respect to the phase, they find

$$k_B T_{hot}^{Kluge} = m_e c^2 \left(\frac{\pi a_0}{2 \log 16 + 2 \log a_0} - 1\right) \tag{2.29}$$

Yet, this scaling does not account for plasma properties and is only valid for step-like density profiles. On the contrary, numerical studies indicate that the plasma density and gradient do play an important role [46, 47]. In particular, it was found that shallow plasma gradients can result in increased electron temperatures.

Closely related to the hot electron generation is the vigorously discussed question of laser energy absorption in overdense plasmas. The generation of hot electrons is a prominent example of coupling laser energy into a plasma, and very often is thought to be the dominant absorption channel. Many different mechanisms eventually lead to the generation of high energetic electrons [1, 28]. At relativistic intensities and steep plasmas gradients, the most dominant absorption processes are *$j \times B$ heating* [29] and *Brunel* or *vacuum heating* [48]. Both processes are physically very similar. In the case of oblique incidence, electrons are driven in the electric field of the laser giving rise to the generation of MeV electron bunches at a frequency of ω_L whereas in the case of normal incidence the magnetic term of the Lorentz force dominates and repetitively generates hot electrons at a frequency of $2\omega_L$ (see discussion above). In experiment, both mechanisms most likely contribute to the measured electron distributions, as even under normal incidence the critical surface deforms during the interaction and eventually results in oblique incidence angle at the side wings of the interaction volume. Owing to the vast variety of competing absorption mechanisms, it is difficult to isolate and study a particular process in experiment. Instead, many processes very often contribute to the recorded electron data making the correct interpretation very complex. As of to date, no comprehensive theory exists and thus the physical understanding of laser absorption still remains somewhat unclear.

2.3.2 Relativistic Electron Mirrors from Nanometer Foils

The interaction of an intense laser pulse with solid density plasma has been envisioned as a way to generate relativistic attosecond electron bunches with densities close to solid [49]. In particular, numerous theoretical work has been devoted very recently to the laser–nanofoil interaction at intensities high enough to achieve complete separation of all electrons from the ions using foil thicknesses of only a few nm [50].

Figure 2.6 illustrates the interaction dynamics in this regime, showing a step-like laser pulse with $a_0 = 48$ incident on an ultrathin (effectively 4 nm) foil. The laser pulse acts like a snowplow, drives out all electrons coherently as a single dense electron layer co-moving with the laser field, whereas the ions rest at their initial position owing to their high inertia. The created electron bunch gains energy as it surfs on the electromagnetic wave of the laser and essentially acts as a superparticle following single electron dynamics. Moreover, as the laser field prevails over the electrostatic fields of the plasma, the electron bunch keeps its initial thickness and density over several laser cycles while being accelerated.

To achieve full charge separation, the electric field of the laser has to exceed the electrostatic field arising from the complete separation of all electrons from the ions. Assuming a top-hat laser pulse and a step-like plasma profile with thickness d, we can estimate when the radiation pressure exceeds the electrostatic field pressure such that no force balance can be reached

$$\frac{I}{c} \gg \frac{1}{2}\epsilon_0 E_{es}^2 \tag{2.30}$$

The electrostatic field simply is $E_{es} = en_e d/\epsilon_0$ in the case of complete charge separation. Using Eq. 2.22 and expressing the laser field in normalized units $a_0 =$

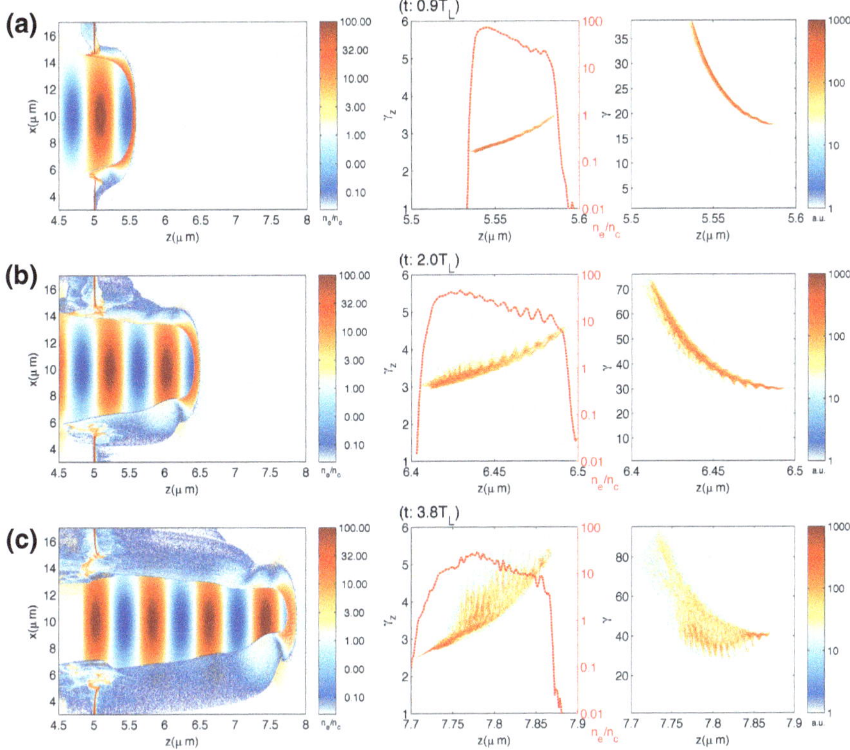

Fig. 2.6 *Laser-driven, relativistic electron mirror from a nanometer foil.* Input parameter: $a_0 = 48.3$ (pulse shape: supergaussian), $Nk_L d = 15.7$ ($N = 100n_c$). Here, $t = 0$ is defined as the timestep when the laser pulse reaches the plasma layer. **a–c** depict different time steps

$eE_0/mc\omega_L$, we can rewrite the electron blowout condition as

$$a_0 \gg \frac{n_e}{n_c} k_L d \tag{2.31}$$

It is worth noting that this condition implies $d/l_s \ll a_0/\sqrt{N}$ with $N = n_e/n_c \gg 1$. Hence, in order to drive out all electrons effectively, the plasma thickness should not be much larger than the skin depth of the laser. Thus, in this scenario, the laser interacts with an overcritical, yet, transparent plasma layer.

This regime was first described by Kulagin et al. [50] and has been investigated in numerous theoretical studies since then [3, 51–53].[6] However, most of this theoretical work relies on highly idealized laser pulses with infinitely steep rise time. Using more realistic pulses with Gaussian rise spanning over many laser cycles [54, 55],

[6] Using a flattop laser pulse profile, the generation of a relativistic electron mirror was studied in great detail in [51] and an empirical lower threshold value $a_{th} = 0.9 + 1.3\,Nk_L d$ was derived from PIC simulations. However, this threshold amplitude is strongly dependent on the temporal laser

the laser nanofoil interaction becomes very complex and yet is very little understood. Advancing this knowledge is the ambition of this thesis.

2.4 Relativistic Doppler Effect

The change in frequency and amplitude of an electromagnetic wave caused by the relative motion of the source and observer was first discussed by Einstein in his work on special relativity [56]. In his paper, Einstein calculates the reflection of an electromagnetic wave from a relativistically fast moving mirror as a working example of Lorentz transformations. The underlying idea is to transform the problem to the rest frame of the mirror, where the reflection of a light wave is well described by basic laws of optics. In the following, we shall briefly repeat Einstein's discussion here, as the result will be an integral part of this thesis.

Let the mirror propagate in $+z$ direction with velocity $\beta = v/c$ and the electromagnetic wave in $-z$ direction with wavevector $k_i = -\omega_L/c$, as shown in Fig. 2.7. As a first step, we transform the incident electromagnetic wave to the rest frame of the mirror making use of the Lorentz boost [2].

$$\begin{aligned} \omega_L'/c &= \gamma\omega_L/c - \gamma\beta k_i = (1+\beta)\gamma\omega_L/c \\ k_i' &= -\gamma\beta\omega_L/c - \gamma k_i = (1+\beta)\gamma k_i \end{aligned}$$

Thus, the incident laser field is blue shifted in the rest frame of the mirror. For the sake of simplicity, we assume a perfect mirror, which reflects back the incident field with $k_r' = -k_i'$. Now, the lab frame moves with $-\beta$ with respect to the rest frame of the mirror. Transforming the reflected light field back to the lab frame, we find

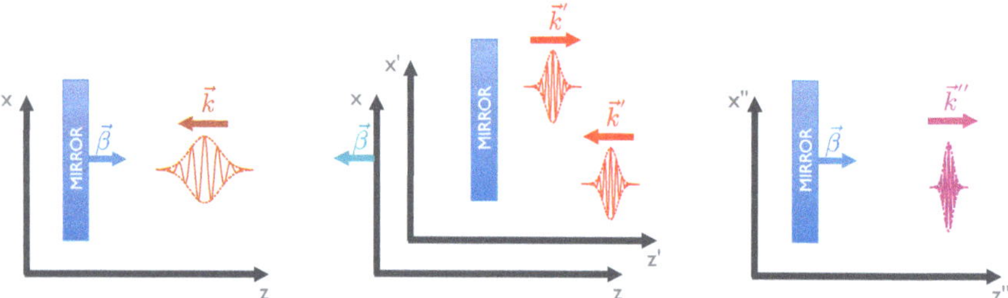

Fig. 2.7 *Relativistic Doppler effect*. Illustration of the Lorentz transformations applied to the system to discuss the reflection of a laser pulse from a counter-propagating mirror, moving with relativistic velocity β

(Footnote 6 continued)
pulse shape that is employed and might be very different for a more realistic few cycle laser pulses with Gaussian rise.

$$\omega_L''/c = \gamma\omega_L'/c - \gamma(-\beta)k_r' = (1+\beta)\gamma\omega_L'/c$$
$$k_r'' = -\gamma(-\beta)\omega_L'/c - \gamma k_r' = (1+\beta)\gamma k_r'$$

Using both equations, we find the prominent result for the reflection of an electromagnetic wave from a moving mirror:

$$\omega_L'' = (1+\beta)^2\gamma^2\omega_L \approx 4\gamma^2\omega_L$$
$$k_r'' = (1+\beta)^2\gamma^2 k_i \approx 4\gamma^2 k_i \qquad (2.32)$$

Apart from the relativistic frequency upshift derived here, the amplitude and the duration of the incident wave are changed accordingly as

$$E'' = (1+\beta)^2\gamma^2 E \qquad (2.33)$$

and

$$\tau'' = \frac{\tau}{(1+\beta)^2\gamma^2} \qquad (2.34)$$

Equation 2.33 is obtained from the Lorentz transformation of the electromagnetic field tensor [2]. The pulse compression (Eq. 2.34) stems from the fact that the phase is an invariant under Lorentz transformations [2].

Thus, for an ideal relativistic mirror, the peak power of the reflected radiation can substantially exceed that of the incident radiation due to the increase in photon energy and accompanying temporal compression.

While theoretically extremely rewarding, the generation of a relativistic structure, with properties sufficient to act as a mirror, is extremely challenging. While electron bunches with very high γ factors can be generated with conventional accelerators, they do not form a reflecting structure analogous to a mirror due to their low density and long bunch duration and therefore the backscattered radiation is incoherent. On the contrary, the interaction of a high intensity laser pulse with a few nanometer thin free-standing foil promises the creation of a solid density, attosecond short electron bunch, which may give access to the coherent regime. In the next sections, we shall develop a deeper, microscopic understanding of the mirror properties of such a unique structure.

2.5 Coherent Thomson Scattering

Light, incident on a charged particle, such as an electron, causes the particle to be accelerated, which in turn emits radiation at the same frequency as the incident electromagnetic wave.[7] This process is referred to as Thomson scattering with cross section $\sigma_T = 6.65 \times 10^{-25}\,\text{cm}^2$ [2].

While the reflection from a mirror is usually discussed quantitatively in the framework of electrodynamics, we shall briefly analyze the reflection process from the perspective of scattering theory, as this directly highlights the main challenges to create a mirror-like structure. In scattering theory, the reflection process is a macroscopic manifestation of scattering occurring on a microscopic level. In that sense, the process is very complex as it requires the coherent behavior of a great number of individual scatterers.

In general, a mirror structure constitutes of a large ensemble of individual scatterers re-emitting light at the interface of two media with a constant phase relation, imposed via the incident light field.

Reflection, i.e. coherent scattering takes place, when many scatterers reside in a volume λ'^3, that is $n'_e\lambda'^3 \gg 1$ [57], where λ' is the wavelength of the incident light and n'_e the electron density, both values evaluated in the rest frame of the mirror.

In this scenario, the distance between adjacent scatterers is significantly shorter than the wavelength of the emitted radiation, thus the relative phases of the interfering wavelets of individual scatterers have to be taken into account to evaluate the resulting field. We shall analyze this in depth in the next section, making use of the formalisms commonly used in scattering theory.

2.5.1 Analytical Model

We start from the Thomson scattering of a single electron. The cross section is defined in such a way that the scattered power is $P_T = \sigma_T I_i$, where I_i is the incident energy flux, i.e. intensity. For an electron bunch, consisting of N scattering electrons, we can deduce the radiated power by summing over the scattering amplitudes of each individual electron while taking into account the relative phase. In general, the spatial phase factor of two scatterers separated by a distance $\mathbf{r}$ is $\phi = \mathbf{q} \cdot \mathbf{r}$, where $\mathbf{q}$ is the momentum transfer or scattering vector [58]. Considering an electron bunch with cross section A, the power incident on the bunch is $P_i = AI_i$. Thus, we can write for the backscattered power

$$P_T = \frac{\sigma_T}{A} \left| \sum_{j=1}^{N} e^{i\mathbf{q}\cdot\mathbf{r_j}} \right|^2 P_i \tag{2.35}$$

The evaluation of this sum is well established in the theory of coherent synchrotron or terahertz radiation [59, 60]. We adapt this method and write

[7] This is true as long as $\hbar\omega \ll m_e c^2$, i.e. as long as the photon recoil $\hbar\omega/c \ll m_e c$ is negligible. Otherwise, the process is described in the framework of Compton scattering.

$$P_T = \frac{\sigma_T}{A}[N[1 - f(\mathbf{q})] + N^2 f(\mathbf{q})]P_i \tag{2.36}$$

where the form factor

$$f(\mathbf{q}) = \left| \int e^{i\mathbf{q}\cdot\mathbf{r}} S(\mathbf{r}) d^3 r \right|^2 \tag{2.37}$$

is the square amplitude of the Fourier transform of the normalized particle distribution function $S(r)$, thus owing to the normalization $f(\mathbf{q}) \leq 1$.

The first term of Eq. 2.36 scales with N and describes the incoherent Thomson scattering, whereas the second term, scaling with N^2 represents the coherent contribution. As N is a large number, typically denoting 10^6–10^8 electrons, the coherent signal enhancement $Nf(k)$ can be huge, making the Thomson scattering in the coherent regime highly efficient.

In the following, we are interested in the coherent signal and define a coherent, or mirror-like reflectivity of the bunch as

$$R_m = \frac{\sigma_T}{A} N^2 f(\mathbf{q}) \tag{2.38}$$

Suppose, the electron bunch density can be modeled as a gaussian with $n_e(z) = n_0 e^{-z^2/d^2}$. Then, the number of electrons contributing to the coherent signal is $N = A \int n_e(z) dz = \sqrt{\pi} A n_0 d$ and we can construct

$$S(z) = \frac{1}{N} A\, n_e(z) = \frac{1}{\sqrt{\pi} d} e^{-z^2/d^2} \tag{2.39}$$

In the backscattering geometry $\mathbf{q} = 2k_L \mathbf{e_z}$ and we find for the form factor of an electron bunch with gaussian bunch shape

$$f(\mathbf{q} = \mathbf{2k_L}) = \left| \int e^{i2k_L z} S(z) dz \right|^2 = e^{-2k^2 d^2} \tag{2.40}$$

Thus, we write for the reflectivity of the electron mirror at rest

$$R_m = \frac{\sigma_T}{A} N^2 e^{-2k^2 d^2} \tag{2.41}$$

Now, considering a mirror moving with relativistic velocity, we transform to the rest frame of the mirror and make use of the previous discussion. In the rest frame of the mirror, the incident light is blue-shifted $k_L' = (1 + \beta)\gamma k_L$ and the electron bunch thickness becomes $d' = \gamma d$. Thus, the mirror reflectivity in the lab frame reads as

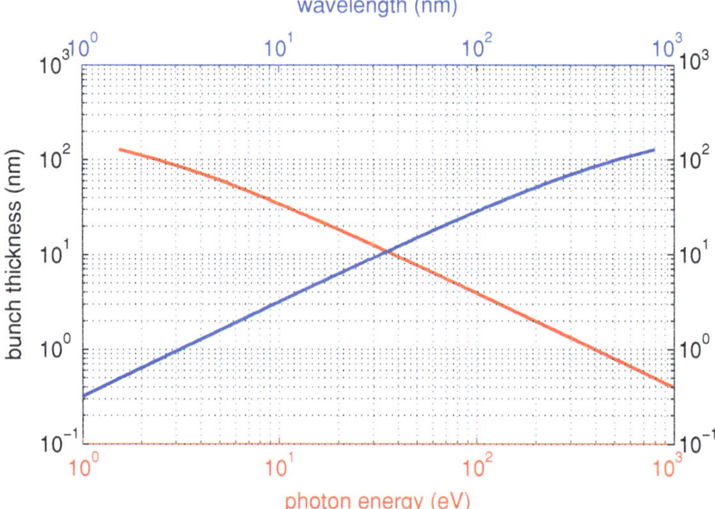

Fig. 2.8 *Dependence of the optimal electron bunch thickness on the upshifted radiation.* Here, the electron bunch thickness is defined by the FWHM value of a gaussian bunch distribution and should not be much larger than $d_{opt} = 1/2k_L\gamma^2$ as the reflectivity rapidly decreases for larger values

$$R_m = \sigma_T \pi A n_0^2 d^2 e^{-2\xi^2}$$
$$\text{with } \xi = (1+\beta)\gamma^2 k_L d \tag{2.42}$$

This formula describes the coherent backscattering from a N electron system. Note, that the coherent enhancement was discussed only in the longitudinal dimension. Thus, the cross-section A in this equation is limited to small values such that the overall quasi-one dimensional geometry of the system is preserved. In more detail, radiation with path length difference of $\Delta > \lambda_r/2$ should not contribute to the coherent enhancement in zeroth order. Thus, an electron located at a distance a from the center, contributes to the signal on axis at a distance R only if $\Delta \sim a^2/R \ll \lambda_r$. In the following, we set $a \sim \lambda_r$, thus $A = \pi\lambda_r^2$.

As an important result of the discussion, we can now define an upper limit on the electron bunch thickness d. Obviously, in order to achieve high reflectivity of the mirror structure $\xi \lesssim 1$, thus

$$k_L d_{opt} \lesssim 1/2\gamma^2. \tag{2.43}$$

It is important to note that not only the length scale (Fig. 2.8), but also the exact shape of the electron distribution is crucial for the bunch reflectivity. Figure 2.9 illustrates that fact by showing the form factor for different bunch shapes and thicknesses. As expected, the form factor drops off more rapidly for shorter wavelengths when increasing the bunch thickness while keeping the bunch shape as a gaussian. However,

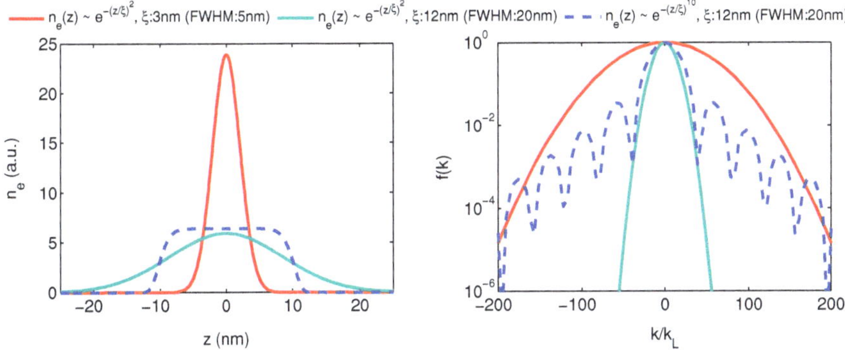

Fig. 2.9 *Dependence of the electron bunch form factor on the electron bunch shape.* The input distribution functions are normalized such that $\int n_e(z)dz = 1$

changing the bunch profile to a steeper, supergaussian profile while keeping the bunch thickness the same, does significantly reduce the fast decay of the form factor for shorter wavelengths. In essence, a mirror structure requires both high density and a sharp mirror to vacuum interface. This implies very steep density gradients, as the length scale of the discontinuity defining the mirror surface needs to be abrupt, well below the wavelength of the emitted light as the backscattered amplitudes would rather cancel out each other in a gradual changing interface [61].

2.5.2 Reflection Coefficients

We can define different reflection coefficients in the case of a moving mirror:

1. the ratio of incident and reflected power

$$R_I = \frac{I_r}{I_i} = (1+\beta)^4\gamma^4 R_m \tag{2.44}$$

2. the ratio of incident and reflected energy, corresponding to the mirror reflectivity of an ordinary mirror

$$R_E = \frac{E_r}{E_i} = \frac{I_r\tau_r}{I_i\tau_i} = (1+\beta)^2\gamma^2 R_m \tag{2.45}$$

   where the underlying assumption is that the mirror lifetime is long compared to the duration of the incident pulse.

3. The ratio of the incident and reflected photon number

$$R_{Phot} = \frac{N_r}{N_i} = \frac{E_r/\hbar\omega_r}{E_i/\hbar\omega_i} = R_E\frac{\omega_i}{\omega_r} = R_m \tag{2.46}$$

2.6 Frequency Upshift from Laser-Driven Relativistic Electron Mirrors

Inherent to the electron motion in a laser field, forward momentum is bound to a transverse momentum, thus each individual electron of the bunch propagates at an angle with respect to the laser axis of the driving laser pulse.

A counter-propagating pulse, incident on such an electron bunch causes each particle to emit dipole radiation.[8] However, as the radiating electron moves at relativistic velocity the emission cone of the radiated field is bent towards the propagation direction of the electron. In consequence, the main contribution of the incoherent signal points off-axis, along the velocity vector $\boldsymbol{\beta}$, as shown in Fig. 2.10.

In contrast, the signal of the coherent scattering is governed by the collective emission of all electrons, which is determined by the interference of the individual backscattered wavelets. Just as in an ordinary reflection, the angle of emission crucially depends in that case on the exact reflection geometry, that is the surface orientation of the scattering structure in connection with the angle of incidence of the impinging laser field and is discussed for arbitrary configurations in [53, 57]. In the counter-propagating geometry, the coherent backscatter signal adds up constructively in mirror surface normal direction, that is in the specular direction, as one would expect intuitively. In contrast, the incoherent signal, points off-axis in bunch velocity direction (Fig. 2.10), and is suppressed by destructive interference. Thus, in the case of coherent backscattering, the frequency upshift is

$$\omega_L'' = (1 + \beta_z)^2 \gamma_z^2 \omega_L \tag{2.47}$$

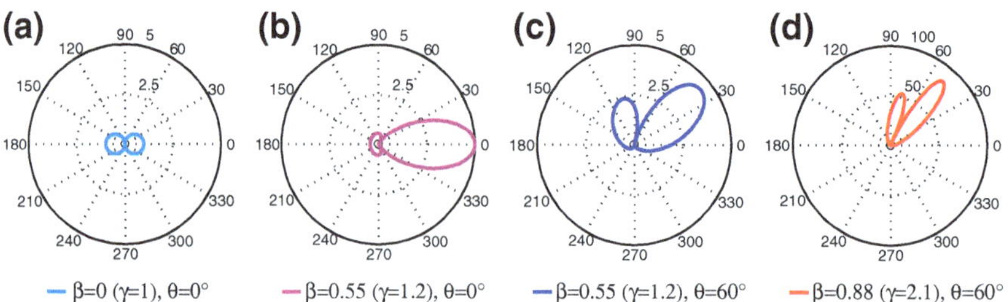

Fig. 2.10 *Dipole emission from a single electron.* Angular dependence of the emitted dipole radiation of an electron propagating with relativistic velocities β in different directions θ, **a–d**

[8] The electric field emitted from a moving charge is (consider field contributions scaling with R^{-1} only):

$$E = \frac{e}{4\pi\epsilon_0 c}\left(\frac{\boldsymbol{n} \times [(\boldsymbol{n} - \boldsymbol{\beta}) \times \dot{\boldsymbol{\beta}}]}{R(1 - \boldsymbol{n} \cdot \boldsymbol{\beta})^3}\right).$$

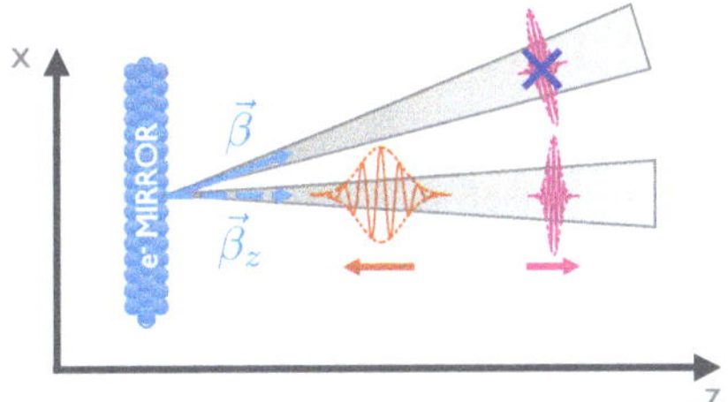

Fig. 2.11 *Relativistic Doppler upshift from laser-driven electron mirrors.* The incoherent signal points close to the direction of $\boldsymbol{\beta}$ and is fully suppressed by destructive interference in the case of a mirror-like reflection. In contrast, the coherent signal is emitted in the direction of specular reflection. Thus, the corresponding velocity component β_z governs the relativistic Doppler upshift $\sim 4\gamma_z^2$

where γ_z is the effect γ factor of the mirror motion in mirror normal direction[9] (Fig. 2.11)

$$\gamma_z = \frac{1}{\sqrt{1-\beta_z^2}} = \frac{\gamma}{\sqrt{1+(p_\perp/mc)^2}}. \tag{2.48}$$

As $p_\perp$ tends to be large due to the transverse field character of the driving laser pulse, γ_z can be significantly smaller than the full γ factor.

References

1. Gibbon P (2005) Short pulse laser interactions with matter: an introduction. Imperial College Press, London
2. Jackson JD (1998) Classical electrodynamics, 3rd edn. Wiley, New York
3. Meyer-ter Vehn J, Wu H-C (2009) Coherent thomson backscattering from laser-driven relativistic ultra-thin electron layers. Eur Phys J D: At Mol Opt Plasma Phys 55:433–441
4. Popov KI (2009) Laser based acceleration of charged particles. PhD thesis, University of Alberta
5. Meyer-ter Vehn J, Pukhov A Relativistic laser plasma interaction: analytical tools. Lecture Notes
6. Landau LD, Lifshitz EM (1980) The classical theory of fields, 4th edn. Course of theoretical physics series, vol 2. Butterworth-Heinemann, Oxford
7. Lawson JD (1979) Lasers and accelerators. IEEE Trans Nucl Sci 26(3):4217–4219
8. Woodward PM (1947) A method of calculating the field over a plane. J Inst Electr Eng 93:1554–1558
9. Palmer RB (1987) An introduction to acceleration mechanisms. SLAC-PUB-4320
10. Esarey E, Sprangle P, Krall J (1995) Laser acceleration of electrons in vacuum. Phys Rev E 52(5):5443–5453

[9] $\gamma^2 = 1 + (p_z/m_ec)^2 + (p_\perp/m_ec)^2 = 1 + (\gamma\beta_z)^2 + (p_\perp/m_ec)^2 \Rightarrow 1 + (p_\perp/m_ec)^2 = \gamma^2(1-\beta_z^2) = \gamma^2\gamma_z^{-2}$.

11. Plettner T, Byer RL, Colby E, Cowan B, Sears CMS, Spencer JE, Siemann RH (2005) Visible-laser acceleration of relativistic electrons in a semi-infinite vacuum. Phys Rev Lett 95(13):134801
12. Dodin IY, Fisch NJ (2003) Relativistic electron acceleration in focused laser fields after above-threshold ionization. Phys Rev E 68:056402
13. Hu SX, Starace AF (2002) Gev electrons from ultraintense laser interaction with highly charged ions. Phys Rev Lett 88(24):245003
14. Hu SX, Starace AF (2006) Laser acceleration of electrons to giga-electron-volt energies using highly charged ions. Phys Rev E 73(6):066502
15. Bauer D, Mulser P, Steeb WH (1995) Relativistic ponderomotive force, uphill acceleration, and transition to chaos. Phys Rev Lett 75(25):4622–4625
16. Quesnel B, Mora P (1998) Theory and simulation of the interaction of ultraintense laser pulses with electrons in vacuum. Phys Rev E 58(3):3719–3732
17. Moore CI, Knauer JP, Meyerhofer DD (1995) Observation of the transition from Thomson to Compton scattering in multiphoton interactions with low-energy electrons. Phys Rev Lett 74(13):2439–2442
18. Meyerhofer DD (1997) High-intensity-laser-electron scattering. IEEE J Quantum Electron 33(11):1935–1941
19. Hartemann FV, Fochs SN, Le Sage GP, Luhmann NC, Woodworth JG, Perry MD, Chen YJ, Kerman AK (1995) Nonlinear ponderomotive scattering of relativistic electrons by an intense laser field at focus. Phys Rev E 51(5):4833–4843
20. Malka G, Lefebvre E, Miquel JL (1997) Experimental observation of electrons accelerated in vacuum to relativistic energies by a high-intensity laser. Phys Rev Lett 78(17):3314–3317
21. McDonald Kirk T (1998) Comment on "experimental observation of electrons accelerated in vacuum to relativistic energies by a high-intensity laser". Phys Rev Lett 80(6):1350
22. Popov KI, Bychenkov VY, Rozmus W, Sydora RD (2008) Electron vacuum acceleration by a tightly focused laser pulse. Phys Plasmas 15(1):013108
23. Wang PX, Ho YK, Yuan XQ, Kong Q, Cao N, Sessler AM, Esarey E, Nishida Y (2001) Vacuum electron acceleration by an intense laser. Appl Phys Lett 78(15):2253–2255
24. Wang PX, Ho YK, Yuan XQ, Kong Q, Cao N, Shao L, Sessler AM, Esarey E, Moshkovich E, Nishida Y, Yugami N, Ito H, Wang JX, Scheid S (2002) Characteristics of laser-driven electron acceleration in vacuum. J Appl Phys 91(2):856–866
25. Pang J, Ho YK, Yuan XQ, Cao N, Kong Q, Wang PX, Shao L, Esarey EH, Sessler AM (2002) Subluminous phase velocity of a focused laser beam and vacuum laser acceleration. Phys Rev E 66(6):066501
26. Chaloupka JL, Meyerhofer DD (1999) Observation of electron trapping in an intense laser beam. Phys Rev Lett 83(22):4538–4541
27. Stupakov GV, Zolotorev MS (2001) Ponderomotive laser acceleration and focusing in vacuum for generation of attosecond electron bunches. Phys Rev Lett 86(23):5274–5277
28. Kruer WL (2003) The physics of laser plasma interactions. Westview Press, Boulder
29. Kruer WL, Estabrook K (1985) J x b heating by very intense laser light. Phys Fluids 28(1):430–432
30. Macchi A (2011) An introduction to ultraintense laser-plasma interactions
31. Rykovanov SG (20029) Interaction of intense laser pulses with overdense plasmas. Theoretical and numerical study. PhD thesis, Ludwig-Maximilians-Universität München (LMU)
32. Baton SD, Santos JJ, Amiranoff F, Popescu H, Gremillet L, Koenig M, Martinolli E, Guilbaud O, Rousseaux C, Rabec Le Gloahec M, Hall T, Batani D, Perelli E, Scianitti F, Cowan TE (2003) Evidence of ultrashort electron bunches in laser-plasma interactions at relativistic intensities. Phys Rev Lett 91:105001
33. Zheng J, Tanaka KA, Sato T, Yabuuchi T, Kurahashi T, Kitagawa Y, Kodama R, Norimatsu T, Yamanaka T (2004) Study of hot electrons by measurement of optical emission from the rear surface of a metallic foil irradiated with ultraintense laser pulse. Phys Rev Lett 92:165001
34. Bezzerides B, Gitomer SJ, Forslund DW (1980) Randomness, maxwellian distributions, and resonance absorption. Phys Rev Lett 44:651–654

35. Gitomer SJ, Jones RD, Begay F, Ehler AW, Kephart JF, Kristal R (1986) Fast ions and hot electrons in the laser-plasma interaction. Phys Fluids 29(8):2679–2688
36. Beg FN, Bell AR, Dangor AE, Danson CN, Fews AP, Glinsky ME, Hammel BA, Lee P, Norreys PA, Tatarakis M (1997) A study of picosecond laser-solid interactions up to 10[sup 19] w cm[sup - 2]. Phys Plasmas 4(2):447–457
37. Antici P, Fuchs J, Borghesi M, Gremillet L, Grismayer T, Sentoku Y, d'Humières E, Cecchetti CA, Mančić A, Pipahl AC, Toncian T, Willi O, Mora P, Audebert P (2008) Hot and cold electron dynamics following high-intensity laser matter interaction. Phys Rev Lett 101(10):105004
38. Chen H, Wilks SC, Kruer WL, Patel PK, Shepherd R (2009) Hot electron energy distributions from ultraintense laser solid interactions. Phys Plasmas 16(2):020705
39. Tanimoto T, Nishiuchi M, Mishima Y, Kikuyama K, Morioka T, Morita K, Kanasaki M, Pirozhkov AS, Yogo A, Ogura K, Fukuda Y, Sakaki H, Sagisaka A, Habara H, Tanaka KA, Kondo K (2012) Electron energy transport in the thin foil driven by high contrast high intensity laser pulse. AIP Conf Proc 1465(1):148–151
40. Mulser P, Bauer D, Ruhl H (2008) Collisionless laser-energy conversion by anharmonic resonance. Phys Rev Lett 101:225002
41. Wilks SC, Kruer WL, Tabak M, Langdon AB (1992) Absorption of ultra-intense laser pulses. Phys Rev Lett 69(9):1383–1386
42. Sherlock M (2009) Universal scaling of the electron distribution function in one-dimensional simulations of relativistic laser-plasma interactions. Phys Plasmas 16(10):103101
43. Haines MG, Wei MS, Beg FN, Stephens RB (2009) Hot-electron temperature and laser-light absorption in fast ignition. Phys Rev Lett 102(4):045008
44. Kluge T, Cowan T, Debus A, Schramm U, Zeil K, Bussmann M (2011) Electron temperature scaling in laser interaction with solids. Phys Rev Lett 107:205003
45. Malka G, Miquel JL (1996) Experimental confirmation of ponderomotive-force electrons produced by an ultrarelativistic laser pulse on a solid target. Phys Rev Lett 77(1):75–78
46. Lefebvre E, Bonnaud G (1997) Nonlinear electron heating in ultrahigh-intensity-laser-plasma interaction. Phys Rev E 55(1):1011–1014
47. Kemp AJ, Sentoku Y, Tabak M (2008) Hot-electron energy coupling in ultraintense laser-matter interaction. Phys Rev Lett 101(7):075004
48. Brunel F (1987) Not-so-resonant, resonant absorption. Phys Rev Lett 59(1):52–55
49. Naumova N, Sokolov I, Nees J, Maksimchuk A, Yanovsky V, Mourou G (2004) Attosecond electron bunches. Phys Rev Lett 93(19):195003
50. Kulagin VV, Cherepenin VA, Hur MS, Suk H (2007) Theoretical investigation of controlled generation of a dense attosecond relativistic electron bunch from the interaction of an ultrashort laser pulse with a nanofilm. Phys Rev Lett 99(12):124801
51. Kulagin VV, Cherepenin VA, Gulyaev YV, Kornienko VN, Pae KH, Valuev VV, Lee J, Suk H (2009) Characteristics of relativistic electron mirrors generated by an ultrashort nonadiabatic laser pulse from a nanofilm. Phys Rev E 80(1):016404
52. Qiao B, Zepf M, Borghesi M, Dromey B, Geissler M (2009) Coherent x-ray production via pulse reflection from laser-driven dense electron sheets. New J Phys 11(10):103042 (11pp)
53. Habs D, Hegelich M, Schreiber J, Gross M, Henig A, Kiefer D, Jung D (2008) Dense laser-driven electron sheets as relativistic mirrors for coherent production of brilliant x-ray and γ-ray beams. Appl Phys B: Lasers Opt 93:349–354
54. Tian Y, Yu W, Lu P, Xu H (2008) Generation of periodic ultrashort electron bunches and strongly asymmetric ion coulomb explosion in nanometer foils interacting with ultra-intense laser pulse. Phys Plasmas 15(5):053105
55. Popov KI, Bychenkov VY, Rozmus W, Sydora RD, Bulanov SS (2009) Vacuum electron acceleration by tightly focused laser pulses with nanoscale targets. Phys Plasmas 16(5):053106
56. Einstein A (2005) Zur Elektrodynamik bewegter Körper [AdP 17, 891 (1905)]. Ann Phys 14(S1):194–224
57. Wu H-C, Meyer-ter Vehn J (2009) The reflectivity of relativistic ultra-thin electron layers. Eur Phys J D: At Mol Opt Plasma Phys 55:443–449
58. Als-Nielsen J, McMorrow D (2001) Elements of modern X-ray physics. Wiley, New York

59. Hirschmugl CJ, Sagurton M, Williams GP (1991) Multiparticle coherence calculations for synchrotron-radiation emission. Phys Rev A 44:1316–1320
60. Carr GL, Martin MC, McKinney WR, Jordan K, Neil GR, Williams GP (2002) High-power terahertz radiation from relativistic electrons. Nature 420(6912):153–156
61. Hecht E (2001) Optics, 4th edn. Addison Wesley, Reading

Chapter 3
Experimental Methods: Lasers, Targets and Detectors

3.1 High Power Laser Systems

High power laser systems are based on the concept of chirped pulse amplification (CPA). This technique was first successfully demonstrated for laser pulses by Strickland and Mourou [1] in 1985, and nowadays is used to amplify ultrashort laser pulses up to the petawatt level. The underlying concept of a CPA laser architecture is shown schematically in Fig. 3.1.

To reach high peak powers, an ultrashort (10–20 fs), low energetic ($\sim$nJ) laser pulse seeded from a mode locked oscillator is amplified in energy by more than 10^8 (or more) orders of magnitude. However, owing to the incredibly high gain, the intensity of the pulse would inevitably exceed the damage threshold of the optical components unless being reduced by either increasing the beam diameter or the duration of the laser pulse. While increasing the beam diameter seems rather simple, it imposes a considerable increase in size and cost upon the system due to the use of large aperture optics. Moreover, it requires crystal sizes clearly beyond those currently available. On the contrary, stretching the pulse temporally prior to the amplification and restoring the initial pulse duration by subsequent re-compression allows for small beam diameters on the gain media.

To stretch the pulse, spectral components of different frequency are set to different beam paths with the aid of dispersive optics, resulting in a temporal elongation of the pulse typically on the order of 100 s ps, with a linearly increasing instantaneous frequency (chirp). After amplification, the frequency chirp is compensated by the grating setup of the compressor, which is set up in vacuum and uses an expanded beam to avoid nonlinearities or even optical damages caused by the drastic increase in intensity as the pulse gets compressed. Brief descriptions of different stretcher and compressor setups can be found here [2].

The amplification of the pulse is carried out in conventional systems by the use of an active medium. Here, the bandwidth of the gain material determines the pulse duration that can be realized. Nowadays, Ti:Sapphire is commonly used due to its broad amplification bandwidth, good heat conductivity, and broad absorption bands

© Springer International Publishing Switzerland 2015

D. Kiefer, *Relativistic Electron Mirrors*, Springer Theses,
DOI 10.1007/978-3-319-07752-9_3

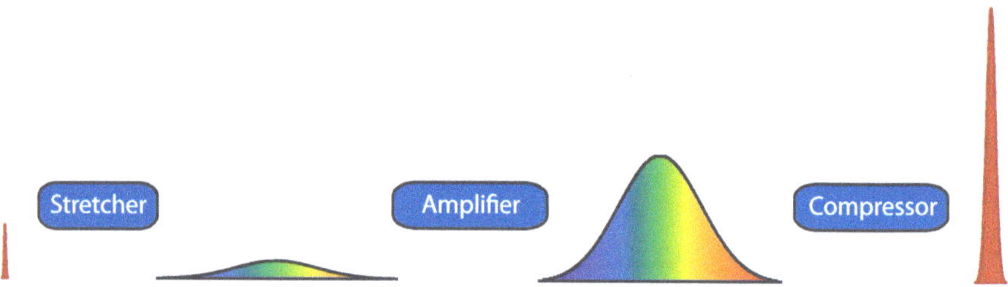

Fig. 3.1 *Chirped pulse amplification (CPA) scheme*. The pulse is stretched by introducing a temporal chirp to reduce the intensity and prevent damages to the optics during the amplification process

in the visible making it suitable for many pump light sources [3]. Ti:Sapphire systems with ~100–20 fs pulse duration, having 1–10 s of joules of pulse energy, reaching peak powers up to the petawatt level have become commercially available and are currently being built all over the world. Pulses of much higher energy (100 s of joules) can be obtained using Nd:Glass as an active medium, which can be produced in large pieces with good optical quality. Here, the gain material is relatively narrow-band and therefore is limited to rather long ~100 s of fs pulse durations.

Ultrashort pulses, close to the single cycle limit can be amplified to high energies using the optical parametric chirped pulse amplification (OPCPA) scheme. Here, instead of using an active medium, the chirped pulse is amplified parametrically in a nonlinear crystal. This scheme is fundamentally different from the conventional laser amplification, as in this nonlinear, three-wave mixing process the energy is directly transferred from the pump to the seed rather than being stored in the active medium. The gain bandwidth is determined by the phase matching condition of the interacting waves and under optimized conditions can extend over a much broader spectral range than in any laser medium. While few cycle pulses are desirable for many interaction schemes, OPCPA laser systems are still in the development phase and to date, only one system exists reaching relativistic intensities in experiments [4].

3.1.1 Laser Pulse Contrast

Apart from the ultrashort, femtosecond pulse duration, high intensity lasers reveal complex time structures on much longer time scales, which is referred to as the laser pulse contrast and turns out to be the key parameter for the use of a laser system in the experiments presented in this thesis. The contrast of a laser pulse is defined as the ratio of the peak intensity to intensity at a given time t and is determined by a great variety of different processes, depicted schematically in Fig. 3.2.

On the nanosecond time scale, a rather constant background noise is typically observed from the amplified spontaneous emission (ASE). Here, luminescent signal owing to spontaneous laser level transitions during the pumping of the gain medium

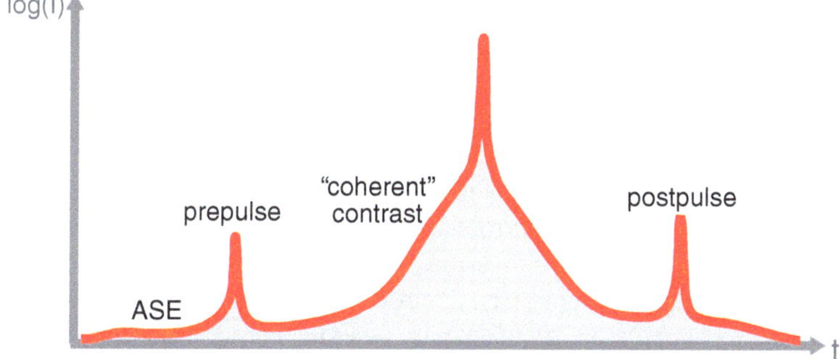

Fig. 3.2 *Illustration of a typical laser pulse contrast curve*. Three distinct features of the temporal contrast are shown: ASE background, various pre- and post pulses and the coherent contrast pedestal at the foot of the main pulse

builds up in the amplifier chain. This process is inherent to the laser amplification, with time scales on the order of the pump duration (ps–ns) and can partially be reduced by optimizing the amplification stages [3]. In addition, discrete pre- or post-pulses can arise on the picosecond to nanosecond time scale. Post-pulses typically arise from multiple reflections in reflective or transmissive optics in the laser system and can in principle be ruled out. Prominent examples are the use of uncoated wave-plates or the confusion of the front and back surface of a dielectric mirror. While at first sight, the suppression of post-pulses seems dispensable, nonlinear coupling of the stretched pulse with its delayed (post-pulse) replica in the gain medium can give rise to a spectral phase modulation, which after the pulse compression results in the formation of a pre-pulse, and thereby degrades the laser pulse contrast considerably [5]. The third and probably least understood characteristic is the exponentially rising pedestal on the tens of picoseconds time scale, referred to as the coherent contrast [6]. This feature frequently observed at the foot of the main pulse usually rises to intensity levels well above the ionization threshold and therefore effectively extends the leading edge of the pulse by a few picoseconds. Recent studies suggest that these incompressible wings of the pulse are due to scatter from the diffraction gratings in the laser pulse stretcher. Reducing the coherent pedestal of CPA high power lasers will be a major challenge over the next years and is essential for the experimental use of future laser systems with envisioned peak powers beyond the petawatt level.

Ultrahigh contrast laser pulses are the prerequisite for experiments with solid density plasmas. In fact, the intensity on target should stay well below the ionization threshold ($\sim 10^{12}\,\mathrm{W/cm^2}$) prior to the main pulse to avoid premature ionization and expansion of the target. Thus, the intensity needs to rise by a factor of 10^8 or more in less than a picosecond—an ultrafast leap in intensity, which conventional CPA systems to date are not capable to deliver.

Different pulse cleaning techniques have been developed to enhance the temporal contrast of the CPA systems on the few picosecond time scale. Among those most commonly used is the cross-wave generation in a nonlinear crystal (XPW) [7] or the use of a plasma mirror (PM) (Appendix A). Nonlinear optical pulse cleaning schemes such as the XPW are effectively loss free, offer high repetition rates and can be implemented directly into the laser chain. However, they cannot be applied after the final compression owing to optical damages (or limited crystal sizes) and therefore are implemented at an intermediate energy level (μJ–mJ) in the amplifier chain using an additional pulse compressor and stretcher before and after the pulse filtering (double CPA [8]). In contrast, plasma mirrors have the great advantage that they can be operated after the final pulse compression. In particular, side wings at the foot of the pulse due to spectral phase noise (coherent pedestal) or imperfect re-compression can be efficiently suppressed. As to date, no other technique is capable of providing such high contrast levels in the near vicinity of the main pulse, plasma mirrors are still widely used in solid target experiments despite their obvious drawbacks such as energy loss, and low repetition rate.

3.1.2 Utilized Laser Systems

The experimental work carried out in the framework of this thesis was conducted at various different high power laser systems, which shall be introduced very briefly in the following.

Los Alamos National Laboratory

The Trident laser facility is a Nd:Glass based, three beam laser system located at the Los Alamos National Laboratory (LANL) in the USA [9]. The laser was originally designed for laser fusion studies in 1980s and still offers μs–ns pulses in beam A and B with a variety of different pulse shapes. The third, short pulse beam C, was upgraded over the years and nowadays reaches peak powers up to 200 TW by making use of the CPA technique.

In spring 2008, right after the completion of the latest laser upgrade, the first thin foil experiment was conducted at the Trident laser facility. At that time, the laser pulse contrast was insufficient for nanometer scale targets and thus a double plasma mirror (DPM) was set up in the target chamber right after the focusing off-axis parabolic mirror, to meet the contrast requirements of the experiment (Appendix A, Fig. A.2, [10]).

In proceeding campaigns, a newly developed pulse cleaning scheme [11] based on the optical parametric amplification (OPA) [12] became available, which allowed omitting the DPM setup and therefore approximately doubling the energy on target. To achieve the intensity needed for the nonlinear filtering process ($\sim$GW/cm^2), the pulse cleaner was positioned in between an additional compressor and stretcher (double CPA), at the 250 μJ level. Here, after a total gain of about $\sim 10^5$, the pulse is recompressed and is split into two replicas, which are used as frequency doubled pump and seed signal in a subsequent OPA stage. The thus generated idler at the

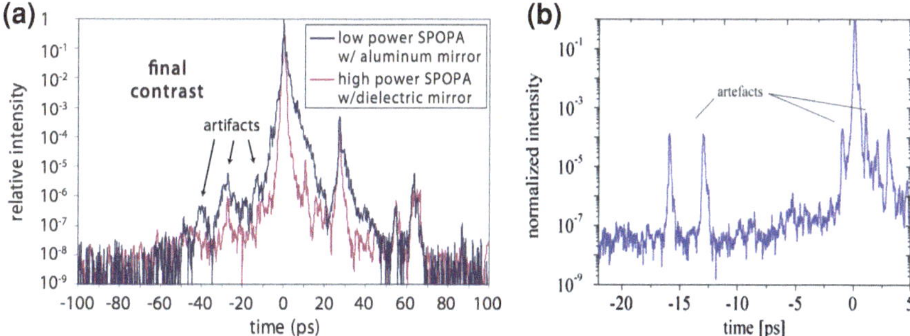

Fig. 3.3 *MBI, LANL laser pulse contrast.* **a** 3rd order cross-correlation (Del Mar Photonics) of the Trident laser pulse using the OPA pulse cleaning front-end (*blue curve*). Positioning the pulse cleaning at later amplification stages and replacing a mirror in the stretcher did increase the final contrast further (*magenta curve*). Due to the low repetition rate of the laser system, the contrast measurement was obtained without firing the final amplifiers, thus could potentially differ from a full power shot on target (by courtesy of R. P. Johnson, LANL). **b** 3rd order cross-correlation of the MBI Ti:Sapphire laser. The contrast on target is further enhanced using a double plasma mirror setup, which is not included in that measurement (by courtesy of S. Steinke, MBI)

fundamental frequency exhibits an inherent ultra-high contrast owing to the short pump pulse duration and is therefore used for further amplification. Moreover, pre-pulses and ASE-pedestal are efficiently suppressed within the amplification window due to the cubic intensity scaling between idler and seed signal. While the idler signal right after the OPA pulse cleaning stage is almost background free [11], the pulse picks up noise as it propagates through the amplifiers in the laser chain. A contrast measurement of the laser pulse taken after the final re-compression is shown in Fig. 3.3a. Since the first implementation of the nonlinear pulse cleaning, the contrast of the laser has been improved further by moving the pulse cleaning to later amplification stages.

Max-Born-Institut

During the MPQ ATLAS laser upgrade, experimental work on the high intensity laser nanofoil interaction was carried out at the Max Born Institute (MBI) in Berlin. The MBI laboratory hosts a 30 TW Ti:Sapphire laser system, which was optimized for high contrast, solid target experiments. A detailed layout of the system is given in [13]. The laser system has a rather good intrinsic contrast ratio of $\sim 10^7$ at -5 ps before the arrival of the main pulse, as can be seen from the autocorrelation measurement, shown in Fig. 3.3b. In addition, a re-collimating double plasma mirror was implemented into the system [14], resulting in an estimated contrast ratio of $\sim 10^{11}$ on the few ps time scale. The system has continuously improved since the beginning of the collaboration and has become a workhorse for laser nanofoil experiments in the recent years.

Rutherford Appleton Laboratory

To date, the Astra Gemini laser is one of the most powerful Ti:Sapphire lasers in the world. It provides two optically synchronized laser pulses, each of which reaching peak powers of up to half of a petawatt [6]. Gemini is an extension of the Astra laser, which served until 2004 and is now used as the input beam for the Gemini system. After the final amplifier of the Astra laser system, the output beam is split into two halves and seed into the Gemini system, which consists of two independent amplification stages including pump lasers and subsequent pulse compressors. After re-compression, both beams are sent to the target chamber independently.

A re-collimating double plasma mirror system was installed in the target chamber, which can be used for either one of the beams to enhance the contrast of the laser pulse [15]. Due to the high contrast requirements needed for thin foil experiments, the contrast of the system was carefully evaluated in the course of the experimental campaign in 2010/2011. Here, as opposed to the contrast measurements presented from LANL and MBI, the full power beam in combination with the double plasma mirror setup was used to obtain the autocorrelation curves (Fig. 3.4). The measurement reveals that the double plasma mirror enhances the contrast ratio by more than four orders of magnitude. As a result, the ionization threshold of the target is reached at around -2 ps prior to the peak of the pulse.

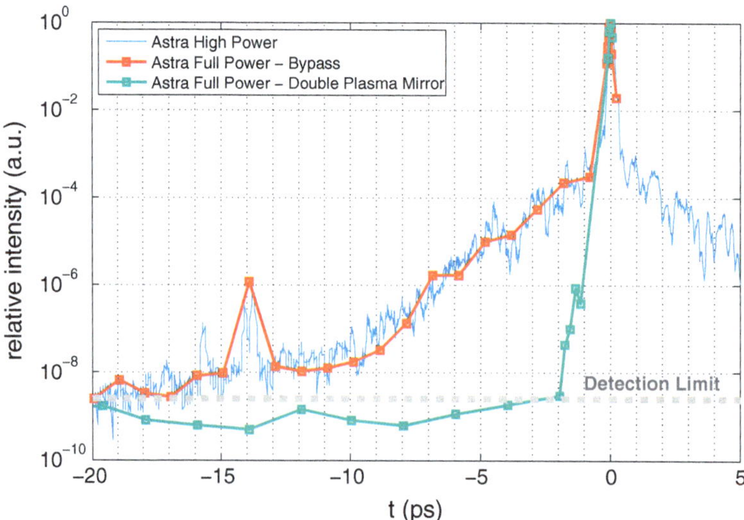

Fig. 3.4 *Astra Gemini laser pulse contrast*. The contrast *curves* were obtained using a scanning third order autocorrelator (Amplitude Sequoia), which was installed after the plasma mirror, next to the target chamber. Astra High Power (*blue curve*), does not use the Gemini amplifiers and therefore operates at 10 Hz, allowing for fast, high resolution scans

Table 3.1 *Summary of the high power laser systems used for experimental studies*

	LANL	MBI	MPQ	ASTRA
Medium	Nd:Glass	Ti:Sapph	Ti:Sapph	Ti:Sapph
Wavelength (μm)	1.053	0.8	0.8	0.8
Rep. rate	1/45 min	10Hz	10Hz	1/min
Energy (J)	80	0.7	0.4	5
Duration (fs)	500	45	30	55
Pulse cleaning	OPA	DPM	DPM	DPM
Peak intensity (W/cm^2)	3×10^{20}	5×10^{19}	8×10^{19}	6×10^{20}
Norm. peak intensity	15	5	6	17

The parameters given here are consistent with the ones seen in the experimental campaigns and my vary slightly from best performance parameters of the systems given elsewhere. The stated energy values refer to the pulse energy on target, taking into account losses from PMs and beamline systems

Max-Planck-Institut of Quantum Optics

At the time ultrathin DLC targets became available at the LMU, the Max-Planck-Institut of Quantum Optics (MPQ) ATLAS laser system reached ~20 TW peak power, but did not have sufficient contrast on the picosecond time scale. The rather quick implementation of a double plasma mirror setup right in front of the thin foil target, analogous to the one successfully used at Trident laser facility (Appendix A), was examined in an experimental study. However, it turned out that this setup is impractical for the ATLAS laser system. The reason is that due to the rather low energy of the ATLAS pulse and the fast focusing parabola (f/3) in the target chamber, the fluence needed to operate a plasma mirror was reached only in the very close vicinity of the target (~1mm) and therefore the PM setup interfered considerably with target alignment and focus diagnostics. In consequence, a re-collimating double plasma mirror system was built, which is presented in great detail in the Appendix A.1.

The parameters of the described laser systems are summarized in Table 3.1

3.2 Diamond-Like Carbon Foils

Carbon exists in a great variety of different amorphous and crystallite structures due to its ability to form atomic bonds in different hybridization states. Most prominent examples are diamond, characterized by its sp^3 hybridized atomic orbitals, and graphite with weaker sp^2 bond configuration.

Diamond-like carbon (DLC) is a meta-stable form of amorphous carbon containing a mixture of sp^3 and sp^2 carbon hybridization states. If a high fraction of sp^3 bonds is reached, one refers to tetrahedral amorphous carbon (ta-C), which recovers many of the extreme properties of diamond such as mechanical hardness and chemical inertness. Moreover, very similar to pure diamond, DLC is optically transparent and owns a wide bandgap.

DLC films have a wide range of applications in industry, where they are mostly used as a protective coating. Owing to the exceptional mechanical properties, DLC films are well-suited for the use as a free-standing foil. A good review article on DLC material discussing the physical properties as well as many of the production and characterization methods is given by [16]. In this chapter, the free-standing DLC foils used for the experimental studies shall be briefly introduced.

Fabrication

Different deposition methods can be employed to produce DLC films [16]. The common feature of all techniques is that the film is formed from a carbon or hydrocarbon ion beam with particle energies on the order of 100 eV. The impact of these energetic ions on a growing film gives rise to the formation of sp^3 bonds—the key component of the DLC material. Depending on the production system, DLC films of various thickness and size can be obtained.

At the LMU, a DLC target laboratory specialized on producing *free-standing*, nm *thin* DLC foils was established over the recent years. Here, DLC films are produced employing a cathodic arc deposition technique [17] and are subsequently attached free-standing to a steel holder making use of a floating technique.

The cathodic arc deposition system relies on a low-voltage, high current plasma discharge. Here, an arc is ignited in a pulsed mode on a graphite cathode, giving rise to the formation of a dense carbon plasma. A fraction of $\sim$10 % of the induced arc current is carried by carbon ions streaming towards the anode with a kinetic energy of $\sim$50 eV, which is controlled by the applied bias voltage. Along with the plasma current, neutral macro-particles are blown off the cathodic spot. To avoid contamination of the DLC film, a 90° magnetic duct is used to filter out neutral particles and guide the carbon ions to the deposition substrate. As a result, a high fraction of sp^3 bonds, up to 75 %, is achieved in the grown film.

The DLC films are deposit on a silicon wafer, which is coated with a thin layer of water soluble NaCl as the release agent. After production, the films are detached from the silicon substrate by immersing them into distilled water, which causes them to release from the wafer and float on the water surface. A steel holder with a regular hole pattern is gently raised from below the floating foil and lift outside the water with the foil stuck to the holder. The film attached to the holder now covers the holes of the target holder free-standing as can be seen in Fig. 3.5.

With these methods, the LMU target fabrication is able to produce free-standing DLC foils with thicknesses ranging from $\sim$60 nm down to $\sim$3 nm and mass density of $\sim$2.7 g/cm^3.

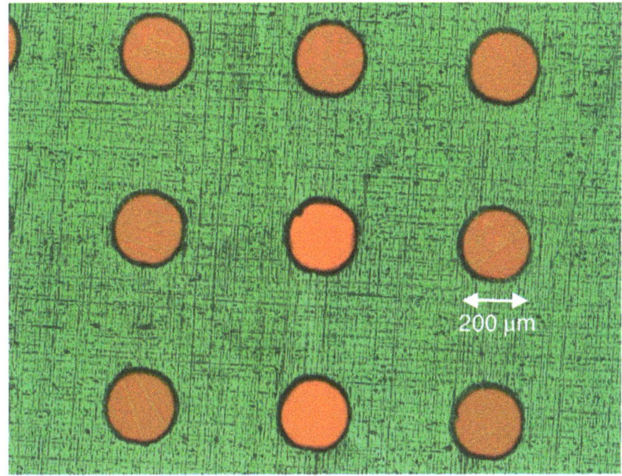

Fig. 3.5 *Microscope image of a sample of free-standing DLC targets*

Characterization

The characterization of extremely thin foils consisting of merely a few atomic layers is challenging. To gain deeper knowledge on the properties of the produced foils, a great variety of different characterization methods have been carried out at the LMU target fabrication laboratory—many of them in close collaboration with different other groups. Most of the employed methods are described in very detail in [18] and therefore shall be given here in brief, only.

A key property with regard to high intensity laser nanofoil experiments is the target thickness, which is determined by the use of an atomic force microscope (AFM). Despite the AFM measurements with sub-nm precision, uncertainties in the actual target thickness persist owing to the fact that the AFM scans are restricted to small areas (tens of μm) and therefore do not resolve the complete thickness topology of the DLC film. Major uncertainties arise from potential inhomogeneities in the ion beam, which introduce thickness gradients ranging over the length scale of the deposition area. To reduce the error, the film is subdivided into six targets and each of those is assigned to an individual thickness deduced from the AFM scan of the corresponding reference sample taken from the close vicinity of each target.

Depending on the quality of the vacuum in the deposition chamber, the produced carbon films can be contaminated with hydrogen ions. To investigate the foil composition in detail, an elastic recoil detection analysis (ERDA) was carried out at the Munich tandem accelerator using a 10 nm thin DLC foil [18]. The ERDA measurement is able to resolve the depth-dependent target composition and revealed a rather constant 10 % hydrogen content throughout the bulk material. Moreover, a $\sim$1 nm thin hydrocarbon contaminant layer was found on the DLC surface.

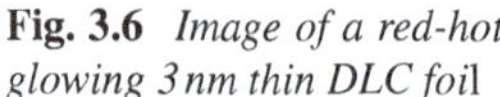

Fig. 3.6 *Image of a red-hot glowing 3 nm thin DLC foil*

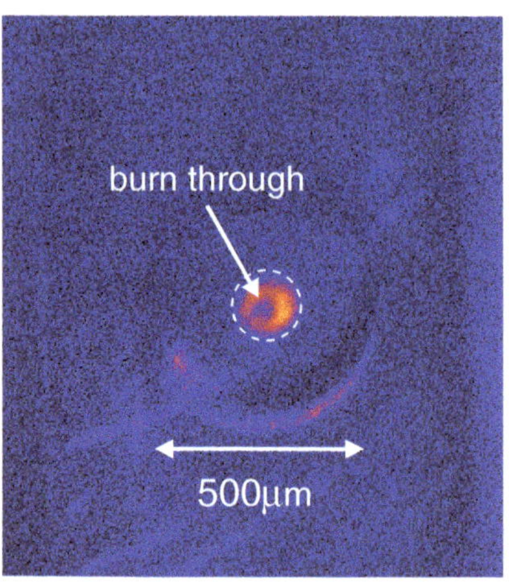

Target Heating

An effective way to remove the hydrogen contaminant layer from the target surface prior to the target shot is to heat up the foil in a clean vacuum environment, e.g. in the evacuated target chamber. The heating process can be carried out by simply irradiating the free-standing foil with a continuous wave (CW) laser. To avoid heating and therefore expansion of the target holder (which could easily cause the breaking of the foil), it is crucial to focus the CW beam carefully to the free-standing foil, exclusively (Fig. 3.6).

As the temperature rises, the hydrogen contaminant layer sublimates from the carbon bulk material, which results in a slight reduction in target thickness. The removal of hydrogen contaminates is evident in the ion signal obtained from full laser shots on pre-heated targets, which showed significantly less to no proton signal in the Thomson parabola spectrometer.

The thermal stability of DLC films was studied in great detail by Kalish et al. [19]. Upon heating, thermally induced relaxation processes can lead to sp^3–sp^2 transformations and clustering of sp^2 domains, which in turn results in the formation of nanocrystallite graphite. However, in the aforementioned study it was found that the thermal stability of the DLC matrix is considerably increased in the case of high sp^3 bonding content. For example, using a DLC film with 80 % sp^3 bonding, no graphitization was observed at a temperature as high as 1270 K.

Laser Damage Threshold

The laser damage threshold of the target material was evaluated in a simple experiment. Here, a 5 nm thin DLC foil was irradiated with the attenuated ATLAS laser pulse, at moderate intensities, close to the ionization threshold. After each shot, the

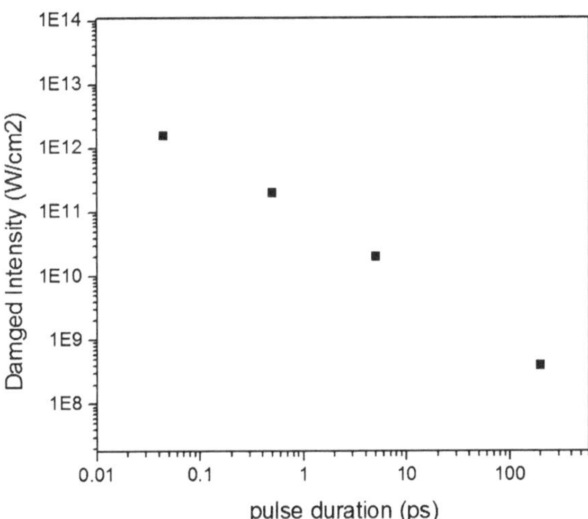

Fig. 3.7 *Target damage threshold intensity* measured for different pulse lengths. By courtesy of W. Ma and J. Bin

foil surface was imaged with the high magnification focal spot diagnostic of the target chamber to identify potential target damage. As the intensity on target is increased, a clear damage (diameter $\sim 10\,\mu$m) is observed at $\sim 2 \times 10^{12}\,\mathrm{W/cm^2}$. The experiment was repeated changing the pulse duration of the ATLAS laser to simulate potential pre-pulses and ASE pedestal (Fig. 3.7).

3.3 Diagnostics

Within the framework of this thesis, various magnetic spectrometers were developed to diagnose the laser–nanofoil interactions. Those spectrometers, which were used in the experimental studies presented in the following chapters, shall be discussed here. First, the underlying concept will be described. Second, the utilized spectrometers will be presented and finally, the employed detectors will be introduced.

3.3.1 Working Principle

A charged particle propagating with kinetic energy E, that enters a uniform magnetic field B with orientation perpendicular to the propagation direction of the particle is forced on a circular orbit with energy dependent radius

$$R = \frac{m_e \gamma \beta}{eB} = \frac{m_e}{eB}\sqrt{\left(1 + \frac{E}{m_e c^2}\right)^2 - 1} \tag{3.1}$$

commonly referred to as the *Larmor radius* [20]. Thus, an electron bunch consisting of a broad energy distribution is dispersed in space via the acting magnetic field. This is the underlying principle of all magnetic spectrometers.

In a rather simple magnet-detector configuration, an analytic expression of the trajectories can be derived directly from Eq. 3.1, which holds true as long as the magnetic field can be treated as a rather idealized constant field. Taking into account magnetic field inhomogeneities, which can become important at the fringe regions of a magnet, additional field components arise and contribute to the particle deflection. While for common magnetic ion detectors (e.g. Thomson parabolas), the fringe fields of a permanent magnet are rather negligible, they do become important for electron measurements owing to the reduced particle mass (factor of 1/1836 or more). Here, the deflection radii caused by additional fringe fields are substantially smaller and thus in general have to be taken into account.

A magnetic field, which accounts for the three-dimensional field distribution of a magnet can be calculated numerically just from the geometry of the magnet using a magnetostatic field solver (CST EM Studio [21]). The numerical results were compared to the actual field distribution deduced from Hall probe measurements many times and generally show excellent agreement to the actual field shape (Appendix B, Fig. B.3). Thus, to obtain the dispersion curve of the spectrometer, monochromatic electron beams of different energy are tracked through the magnet-detector system in a numerical simulation, which solves the equations of motion in the three-dimensional field distribution of the magnet. With the aid of this numerical approach, complex spectrometer configurations of any kind can be treated, which in particular becomes important for rather advanced geometries (Appendix B, Sect. B.1).

3.3.2 Electron Spectrometer

An electron spectrometer was designed to measure the hot electron distribution from laser plasma interactions (Fig. 3.8). As the generated electron beams observed from different laser systems differ fundamentally in their energy distribution, the spectrometer can be equipped with two different magnets optimized for either low (few MeV) or high energetic (several tens of MeV) electrons. In addition, the spectrometer can be operated with either image plate detectors (Sect. 3.3.4) or a scintillator screen (Sect. 3.3.5). While image plates provide high resolution and sensitivity and are in particular suitable for experimental campaigns carried out at low repetition rate Nd:Glass lasers, optical online detection using a scintillator screen in combination with a camera is more appropriate for experiments using Ti:Sapph laser systems, as those systems can be operated at higher repetition rates.

The spectrometer was designed with the aid of numerical simulations (CST), as described in Sect. 3.3.1. The resulting dispersion curves are shown in (Appendix B, Fig. B.4a, b). The magnetic field distributions found from simulations show excellent

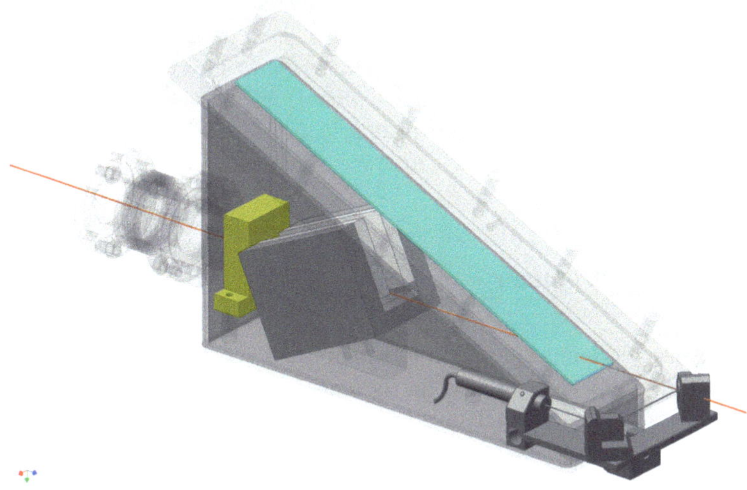

Fig. 3.8 *Electron spectrometer*. The CAD drawing depicts the spectrometer configuration which employs a scintillating screen as detector. When using image plates instead, the acrylic glass cover is replaced by a 1 mm thin Aluminum lid and the image plates are fielded from the outside region, allowing fast detector readout and replacement without breaking the vacuum inside. The spectrometer is equipped with a diode laser to facilitate the spectrometer alignment

agreement to the magnetic field measurements taken from the assembled magnets (Appendix B, Fig. B.3).

The electron signal S recorded on the detector was converted to electron numbers taking into account the detector sensitivity C (Sects. 3.3.4 and 3.3.5). Electron beams incident at an oblique angle on the detector give rise to an enhancement of the detector signal due to the increase in path length $\propto 1/\cos\theta$ and therefore energy deposition. To correct for that, the angle of incidence θ is extracted from the simulation for different detector positions (Appendix B, Fig. B.4) and the detector signal is converted to particle numbers using $N = S\,C\cos\theta$ [22, 23].

To obtain a spectrum from the recorded data, the measured particle trace is subdivided into spatial bins $[x_i, x_i + \Delta x]$, separated by the distance Δx. Each bin corresponds to an energy interval $[E, E + \Delta E]$, which can be deduced from the dispersion curve of the instrument. Owing to the nonlinear energy dispersion, the spectral bandwidth of the energy intervals varies and is determined by the slope of the dispersion curve $\Delta E \sim dE/dx\,\Delta x$. Thus, to calculate the spectrum, the number of particles ΔN within each bin is determined and divided by the spectral bandwidth ΔE of the respective interval.

$$\frac{dN}{dE} \approx \frac{\Delta N}{\Delta E} = \frac{\Delta S}{\Delta E} C \cos\theta \tag{3.2}$$

The bin size Δx is constant and should be chosen not smaller than the size of the pinhole projected onto the detector. Thus, for a given pinhole size D the bin size $\Delta x = (1+M)D/\cos\theta$, where $M = b/a$ is the magnification of the pinhole camera, that is the ratio of the distances a: source—pinhole and b: pinhole—detector.

The spectrometer is versatile tool to record electron distributions from laser plasma experiments and was used for the electron measurements presented in Chap. 4.

3.3.3 Multi-spectrometer

High power laser systems are still in its infancy and very often subject to ongoing research and development. Most of the systems suffer from unstable operation, or at least significant shot-to-shot fluctuations of the laser pulse parameters. This poses grand challenges to experiments with very low repetition rate (or even single shot experiments) as the recorded data very often exhibits significant variations from allegedly identical shots. Apart from improving the laser performance, the best way of tackling this problem is to capture as much information as needed simultaneously, in a single shot.

As part of the Astra Gemini campaign (Chap. 5), a novel Multi-Spectrometer was designed to capture the electron, ion and XUV distribution simultaneously (Fig. 3.9). Figure 3.10 illustrates the setup of the Multi-Spectrometer schematically. The spectrometer essentially consists of three dispersive elements: a magnet, a pair of electric

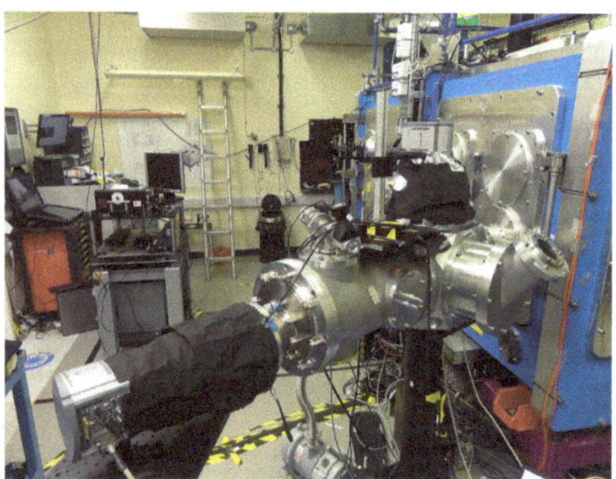

Fig. 3.9 *Photograph of the multi-spectrometer setup.* Electrons are deflected upwards and measured from the top using a scintillator imaged onto a EMCCD camera. The deflection introduced to the ion and photon beam is significantly less and thus both signals are recorded in transmission, at the back of the spectrometer using a MCP detector imaged onto an additional camera. The length *scale* of the spectrometer was about 2 m

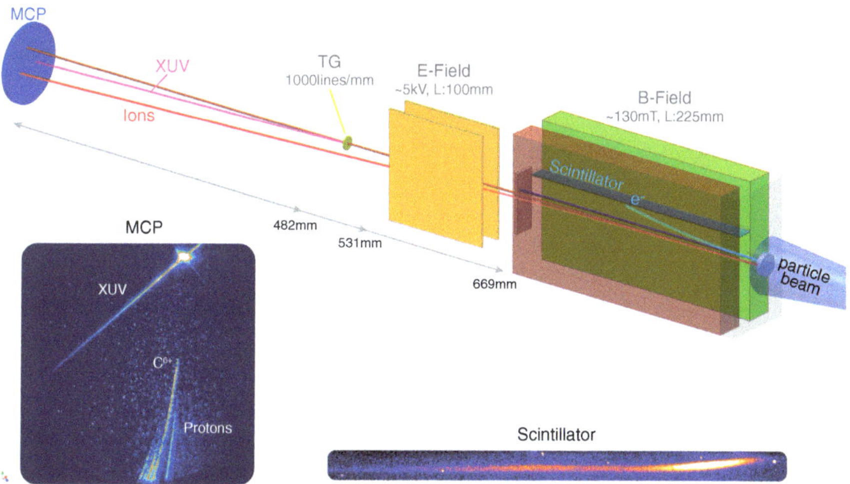

Fig. 3.10 *Setup of the multi-spectrometer*

field plates and a transmission grating combined with a scintillator for electron detection and a micro-channel plate (MCP) recording the ion and photon signal. Due to different sensitivities of the detectors, two separate pinholes of different diameters are used to reach the particle fluxes needed for good signal levels on each detector. To ensure sufficient flux on the scintillating screen a pinhole with diameter D_1: 2 mm is used and a smaller, D_2: 200 µm pinhole is chosen for the MCP detector. The distance in between both pinholes is ~5 mm (or 3.7 mrad with respect to the target), a separation which for most experimental studies is rather negligible.

Permanent magnets typically used in Thomson parabola spectrometers have high magnetic fields ($B \sim 0.5$ T) and hence are too strong for the detection of rather low energetic, few MeV electrons. Therefore, a rather low dispersive magnet with a homogeneous field $B \sim 130$ mT extending over a rather long distance L: 225 mm [24] was employed to resolve electron energies in the range of 1–30 MeV on the scintillator screen. The scintillator was positioned inside the magnet, 15 mm above the entrance pinhole and the resultant electron dispersion on the detector is given in Appendix B, Fig. B.4c.

Magnetic and electric field combined act as a Thomson parabola spectrometer, capable of resolving the energy distributions of different ion species and charge states in a single shot. Here, the magnetic field disperses ions of different energy and the electric field separates ions of different q/m ratios, which in sum gives rise to parabolic ion traces on the detector. Thomsons parabola detectors are widely used in the field and extensively discussed in literature [25–27]. In the Multi-Spectrometer setup, a long drift (669 mm) between the deflecting magnet and the MCP detector is chosen to counterbalance the low ion dispersion in the employed magnetic field.

In order to spectrally resolve the photon signal a transmission grating (TG) was carefully positioned behind the electric field plates at a distance $d = 482$ mm with respect to the MCP detector. At that position, a 50 MeV proton beam is deflected by ~7 mm from the zero axis, and thus does not interfere with the transmission grating located on axis.

Both detectors were imaged with light efficient 50 mm f/1 lenses in combination with low noise EMCCD cameras (Andor: iXon EMCCD). These cameras are capable of amplifying the recorded signal using an integrated electron multiplying (EM) gain, which improves the signal-to-noise ratio and which was used for the detection of electron signal from the scintillator.

The transmission grating implemented into the spectrometer is similar to the ones used for the Low Energy Transmission Grating (LETG) on the Chandra X-ray Observatory satellite [28, 29]. The grating is made out of free-standing gold wires (period: G = 1000 lines/mm), held by two different support mesh structures with lower periodicity. As these supporting line structures act as a grating themselves, they are oriented in directions different from the dispersion axis such that the residual diffraction patterns do not interfere with the dispersed signal. The transmission grating is optimized for 1st order diffraction. The diffraction efficiency in higher orders is reduced by more than one order of magnitude and thus can be neglected [28].

The transmission grating sets an almost linear dispersion $d\lambda/dx \sim 1/Gd \sim$ 2 nm/mm at the detector plane and the wavelength of the recorded radiation can be determined from the interference condition

$$\lambda = \frac{1}{G} \sin\left(\arctan\frac{x}{d}\right) \approx \frac{1}{Gd} x \tag{3.3}$$

The spectral resolution of the spectrometer can be estimated taking into account the imaging properties of the instrument, which can be regarded as a combination of a pinhole camera and a spectrometer. The resultant spot size of the signal S on the detector is estimated from geometrical considerations, which translates to the theoretically expected spectral resolution [30]

$$\Delta\lambda = \frac{d\lambda}{dx} S = \frac{1}{Gd}\left(D + \frac{b}{a}(p + D)\right) \tag{3.4}$$

where p is the source size diameter. Neglecting the source size ($p = 0\,\mu\text{m}$), $\Delta\lambda$ ~0.7 nm, whereas for a rather large source size $p = 200\,\mu\text{m}$, $\Delta\lambda$ equates to ~1 nm. Thus, the spectrometer has good resolution over a broad spectral range (10–100 nm) with $\Delta\lambda/\lambda < 10\,\%$. In practice, the upper limit of the photon energies that can be detected is determined by the saturated zero point, which blurs out to the adjacent short wavelength range. In the experiment presented in Chap. 5, the measurement was limited to wavelengths above >10 nm.

3.3.4 Image Plates

Image plates were developed in the early 1980s for diagnostic radiography as an alternative to conventional X-ray films [31]. They are nowadays routinely used in medical applications as well as in fundamental research such as in X-ray crystallography [32]. Image plates contain an active layer of photostimulable phosphor crystals ($BaFBr : Eu^{2+}$), which is capable of storing a fraction of incident energy and releasing it, when stimulated by visible light.

When ionizing radiation is absorbed in the sensitive phosphor layer, electrons of the Eu^{2+} ions are excited to the conduction band and trapped in color centers of the crystal lattice. These electrons remain in this energetically higher, meta-stable state until exposed to visible or infrared light, which induces the release of the trapped electrons and the decay back to the ground state, which in turn causes the emission of luminescent light (390nm). This process known as photostimulated luminenscence (PSL) is in proportion to the number of trapped electrons and therefore proportional to the incident radiation.

Image plates are read out after exposure with the use of commercial image plate scanners, which stimulate the image plate with a HeNe laser (633 nm) and detect the luminescent signal with a photomultiplier tube (PMT). The output of the PMT is logarithmically amplified and stored as a digital image. Before proceeding with any data analysis, the logarithmic signal needs to be converted to linear PSL values, which can be done using the conversion formula [33]:

$$\mathrm{PSL} = \left(\frac{\mathrm{res}}{100}\right)^2 \frac{4000}{\mathrm{S}} 10^{\mathrm{L}\left(\frac{\mathrm{QL}}{65535}-0.5\right)} \tag{3.5}$$

with scan parameters S: sensitivity, res: scan resolution, L: lattitude and QL: quantum level (raw signal on logarithmic scale). After readout, the residual image stored on the image plate can be erased completely through further illumination to white light, allowing the image plate to be reused many times for data acquisition.

Image plates feature desirable detector characteristics such as high sensitivity, high dynamic range ($\sim 10^5$) and high resolution ($\sim 25\,\mu$m). Moreover, they are resistant to strong electromagnetic pulse (EMP) noise, which is typical for high intensity laser plasma interactions and frequently causes problems when using sensitive electronic devices such as cameras or controllers. To date, image plates have proven as a versatile detector in laser plasma experiments and their response to electron [23, 34, 35], ion [36] and X-ray [37, 38] beams has been studied in great detail.

In this work, electron spectrometers were equipped with image plates as a detector and the calibrations given in [23, 34] were employed to convert the recorded signal to particle numbers. Image plates of type BAS-SR and BAS-TR (FujiFilm) were used in combination with the image plate scanner FujiFilm FLA-7000. The sensitivity to high energetic electrons was found to be almost constant for electron energies above $\sim$1 MeV [23] with $\sim$0.01 PSL/electron in the case of the image plate BAS-SR. For

BAS-TR, this value is reduced by a factor ~3 owing to different thicknesses and densities of the active layer [34].

3.3.5 Scintillators

Scintillating phosphor screens imaged onto a CCD camera offer high repetition, online detection and are nowadays widely used for the detection of electron beams in laser wakefield acceleration (LWFA) experiments. As part of this thesis, scintillators were introduced to solid density plasma experiments as an alternative to the low repetition image plates commonly used in these experiments. Unlike the well-collimated, quasi-monoenergetic electron beams from LWFA, electron distributions from solid density plasmas typically have large divergence angles corresponding to low electron fluxes at the location of the electron spectrometer. Thus, the detection of electrons is crucially dependent on the efficiency of the utilized screen and therefore the most sensitive screen (Kodak Biomax MS, [39]) was used in the experiments. This screen emits in the visible (546 nm) and was imaged with a light efficient objective lens on a low noise CCD camera using shutter times (10–50 ms) much longer than the decay time of the scintillating screen (~1 ms).

When transferring the recorded signal to particle numbers the collection efficiency of the optical imaging needs to be evaluated. To avoid absolute re-calibration of the optical imaging system after every change in the setup, the scintillator signal was referenced to a constant light source (scintillating tritium-filled capsule, mb-microtec) that was cross-calibrated to the response of the scintillator in a previous study [39] using a well-defined high energetic electron beam. In experiment, the constant light source was placed directly on the scintillating screen next to the electron signal and both signals were recorded simultaneously. The direct comparison of the electron signal to the signal intensity of the calibrated light source allowed for the conversion to electron numbers.

Simulation and experiments [40, 41] show that the energy deposition in the screen can be assumed to be constant for electron energies above 1–3 MeV. The onset of this plateau region depends on the exact layer composition of the screen and therefore may vary slightly for different types of screens. In the experiments, a clear departure from the exponential shape of the hot electron distributions was observed at electron energies below ~1 MeV, which was ascribed to the expected energy dependent response of the detector at the low energy end.

References

1. Strickland D, Mourou G (1985) Compression of amplified chirped optical pulses. Opt Commun 55(6):447–449
2. Pretzler G (2000) Höchstleistungs-Kurzpulslaser

3. Koechner W (2006) Solid-state laser engineering. Springer series in optical sciences, 6th revised and updated edn. Springer, Berlin
4. Herrmann D, Veisz L, Tautz R, Tavella F, Schmid K, Pervak V, Krausz F (2009) Generation of sub-three-cycle, 16 tw light pulses by using noncollinear optical parametric chirped-pulse amplification. Opt Lett 34(16):2459–2461
5. Didenko NV, Konyashchenko AV, Lutsenko AP, Tenyakov SY (2008) Contrast degradation in a chirped-pulse amplifier due to generation of prepulses by postpulses. Opt Express 16(5):3178–3190
6. Hooker C, Tang Y, Chekhlov O, Collier J, Divall E, Ertel K, Hawkes S, Parry B, Rajeev PP (2011) Improving coherent contrast of petawatt laser pulses. Opt. Express 19(3):2193–2203
7. Jullien A, Albert O, Burgy F, Hamoniaux G, Rousseau J-P, Chambaret J-P, Augé-Rochereau F, Chériaux G, Etchepare J, Minkovski N, Saltiel SM (2005) 10^{-10} temporal contrast for femtosecond ultraintense lasers by cross-polarized wave generation. Opt Lett 30(8):920–922
8. Kalashnikov MP, Risse E, Schönnagel H, Sandner W (2005) Double chirped-pulse-amplification laser: a way to clean pulses temporally. Opt Lett 30(8):923–925
9. Batha SH, Aragonez R, Archuleta FL, Archuleta TN, Benage JF, Cobble JA, Cowan JS, Fatherley VE, Flippo KA, Gautier DC, Gonzales RP, Greenfield SR, Hegelich BM, Hurry TR, Johnson RP, Kline JL, Letzring SA, Loomis EN, Lopez FE, Luo SN, Montgomery DS, Oertel JA, Paisley DL, Reid SM, Sanchez PG, Seifter A, Shimada T, Workman JB (2008) Trident high-energy-density facility experimental capabilities and diagnostics. Rev Sci Instrum 79(10):10F305
10. Henig A, Kiefer D, Markey K, Gautier DC, Flippo KA, Letzring S, Johnson RP, Shimada T, Yin L, Albright BJ, Bowers KJ, Fernández JC, Rykovanov SG, Wu H-C, Zepf M, Jung D, Liechtenstein VKh, Schreiber J, Habs D, Hegelich BM (2009) Enhanced laser-driven ion acceleration in the relativistic transparency regime. Phys Rev Lett 103(4):045002
11. Shah RC, Johnson RP, Shimada T, Flippo KA, Fernandez JC, Hegelich BM (2009) High-temporal contrast using low-gain optical parametric amplification. Opt Lett 34(15):2273–2275
12. Boyd RW (2008) Nonlinear optics, 3rd edn. Academic Press, Burlington
13. Sokollik T (2008) Investigation of field dynamics in laser plasmas with proton imaging. Ph.D. thesis, Technische Universität Berlin
14. Steinke S (2010) Ion accelertion in the laser transparency regime. Ph.D. thesis, Technische Universität Berlin
15. Streeter MJV, Foster PS, Cameron FH, Bickerton R, Blake S, Brummit P, Costello B, Divall E, Hooker C, Holligan P, Neville DR, Rajeev PP, Rose D, Suarez-Merchen J, Neely D, Carroll DC, Romangnani L, Borghesi M (2009) Astra gemini compact plasma mirror system. Central Laser Facility Annual Report
16. Robertson J (2002) Diamond-like amorphous carbon. Mater Sci Eng R 37:129–281
17. Brown IG (1998) Cathodic arc deposition of films. Annu Rev Mater Sci 28:243–269
18. Henig A (2010) Advanced approaches to high intensity laser-driven ion acceleration. Ph.D. thesis, Ludwig-Maximilians-Universität München (LMU)
19. Kalish R, Lifshitz Y, Nugent K, Prawer S (1999) Thermal stability and relaxation in diamond-like-carbon. A Raman study of films with different sp[sup 3] fractions (ta-c to a-c). Appl Phys Lett 74(20):2936–2938
20. Jackson JD (1998) Classical electrodynamics, 3rd edn. Wiley, New York
21. CST. URL http://www.cst.com/
22. Gotchev OV, Brijesh P, Nilson PM, Stoeckl C, Meyerhofer DD (2008) A compact, multiangle electron spectrometer for ultraintense laser-plasma interaction experiments. Rev Sci Instrum 79(5):053505
23. Tanaka KA, Yabuuchi T, Sato T, Kodama R, Kitagawa Y, Takahashi T, Ikeda T, Honda Y, Okuda S (2005) Calibration of imaging plate for high energy electron spectrometer. Rev Sci Instrum 76(1):013507
24. Gahn C, Tsakiris GD, Witte KJ, Thirolf P, Habs D (2000) A novel 45-channel electron spectrometer for high intensity laser-plasma interaction studies. Rev Sci Instrum 71(4):1642–1645
25. Schreiber J (2006) Ion acceleration driven by high-intensity laser pulses. Ph.D. thesis, Ludwig-Maximilians-Universität München (LMU)

26. Jung D, Hörlein R, Kiefer D, Letzring S, Gautier DC, Schramm U, Hübsch C, Öhm R, Albright BJ, Fernandez JC, Habs D, Hegelich BM (2011) Development of a high resolution and high dispersion thomson parabola. Rev Sci Instrum 82(1):013306
27. Harres K, Schollmeier M, Brambrink E, Audebert P, Blazevic A, Flippo K, Gautier DC, Geissel M, Hegelich BM, Nurnberg F, Schreiber J, Wahl H, Roth M (2008) Development and calibration of a thomson parabola with microchannel plate for the detection of laser-accelerated mev ions. Rev Sci Instrum 79(9):093306
28. Chandra X-ray Observatory. URL http://cxc.harvard.edu/proposer/POG/html/chap9.html
29. Predehl P, Kraus H, Braeuninger HW, Burkert W, Kettenring G, Lochbihler H (1992) Grating elements for the axaf low-energy transmission grating spectrometer. SPIE, vol 1743, pp 475–481
30. Ter-Avetisyan S, Ramakrishna B, Doria D, Sarri G, Zepf M, Borghesi M, Ehrentraut L, Stiel H, Steinke S, Priebe G, Schnürer M, Nickles PV, Sandner W (2009) Complementary ion and extreme ultra-violet spectrometer for laser-plasma diagnosis. Rev Sci Instrum 80(10):103302
31. Iwabuchi Y, Mori N, Takahashi K, Ta M, Shionoya S (1994) Mechanism of photostimulated luminescence process in $BaFBr : Eu^{2+}$ phosphors. Jpn J Appl Phys 33(Part 1, No.1A):178–185
32. Amemiya Y, Miyahara J (1988) Imaging plate illuminates many fields. Nature 336:89–90
33. Paterson IJ, Clarke RJ, Woolsey NC, Gregori G (2008) Image plate response for conditions relevant to laser-plasma interaction experiments. Meas Sci Technol 19(9):095301
34. Hidding B, Pretzler G, Clever M, Brandl F, Zamponi F, Lubcke A, Kampfer T, Uschmann I, Forster E, Schramm U, Sauerbrey R, Kroupp E, Veisz L, Schmid K, Benavides S, Karsch S (2007) Novel method for characterizing relativistic electron beams in a harsh laser-plasma environment. Rev Sci Instrum 78(8):083301
35. Hui C, Hazi AU, van Maren R, Chen SN, Fuchs J, Gauthier M, Le Pape S, Rygg JR, Shepherd R (2010) An imaging proton spectrometer for short-pulse laser plasma experiments. Rev Sci Instrum 81(10):10D314
36. Mancic A, Fuchs J, Antici P, Gaillard SA, Audebert P (2008) Absolute calibration of photo-stimulable image plate detectors used as (0.5-20 mev) high-energy proton detectors. Rev Sci Instrum 79(7):073301
37. Gales SG, Bentley CD (2004) Image plates as X-ray detectors in plasma physics experiments. Rev Sci Instrum 75(10):4001–4003
38. Izumi N, Snavely R, Gregori G, Koch JA, Park H-S, Remington BA (2006) Application of imaging plates to X-ray imaging and spectroscopy in laser plasma experiments (invited). Rev Sci Instrum 77(10):10E325
39. Buck A, Zeil K, Popp A, Schmid K, Jochmann A, Kraft SD, Hidding B, Kudyakov T, Sears CMS, Veisz L, Karsch S, Pawelke J, Sauerbrey R, Cowan T, Krausz F, Schramm U (2010) Absolute charge calibration of scintillating screens for relativistic electron detection. Rev Sci Instrum 81(3):033301
40. Glinec Y, Faure J, Guemnie-Tafo A, Malka V, Monard H, Larbre JP, De Waele V, Marignier JL, Mostafavi M (2006) Absolute calibration for a broad range single shot electron spectrometer. Rev Sci Instrum 77(10):103301
41. Masuda S, Miura E, Koyama K, Kato S (2008) Absolute calibration of an electron spectrometer using high energy electrons produced by the laser-plasma interaction. Rev Sci Instrum 79(8):083301

Chapter 4
Electron Acceleration from Laser–Nanofoil Interactions

While the generation of relativistic electron mirrors from nanoscale foils has attracted great interest, the electron dynamics in high intensity laser nanofoil interactions has not been given great attention, so far. The reason may be due to the well-known complexity of the electron dynamics in laser solid interactions. A better understanding, however, is indispensable for the envisioned application of X-ray generation via Thomson backscattering as well as for the generation of high energetic ion beams.

The difficulty is not only of theoretical nature. To enter the regime of efficient electron mirror generation it is clear that extremely thin, free-standing foils—consisting of only a few atomic layers—are needed. At LMU, great efforts have been made to produce free-standing DLC foils as thin as 3 nm in thickness. Irradiating such a foil with a high contrast laser reaching $a_0 \sim 15$, one would expect to observe the onset of efficient electron blowout as the driving laser field would clearly exceed any restoring electrostatic charge separation field that could build up during the interaction (even in the case of full separation of all electrons from the ions $E_s \sim Nk_L d \sim 10$, Sect. 2.3.2).

The intention of this chapter is to investigate experimentally the electron beams generated in laser-nanometer foil interactions using laser pulse and target parameters available with present day technology. To get first an insight into the dynamics of laser–nanofoil interaction a PIC simulation well-adapted to the experimental configuration is discussed. In the following, experimental data taken from three different laser systems is presented. We observed an increase in the electron mean energies as the target thickness is reduced to the nanometer scale. Quasi-monoenergetic electron beams were observed from ultrathin 3 to 5 nm thin foils using the MBI and LANL laser system.

4.1 PIC Simulation

To elucidate the interaction dynamics of a high intensity laser pulse with a few nanometer thin foil, two dimensional particle-in-cell simulations were carried out using realistic laser pulse parameters. Here, we restrict our computational analysis

© Springer International Publishing Switzerland 2015
D. Kiefer, *Relativistic Electron Mirrors*, Springer Theses,
DOI 10.1007/978-3-319-07752-9_4

to the short pulse, MBI laser system, as for the (factor of ten) longer LANL laser pulse, considerably more computational effort is needed. Hence, in the presented numerical study, we use a laser pulse of 45 fs FWHM duration focused to a 4 μm FWHM spot. The initialized pulse is of Gaussian shape both in space and time and reaches a peak field of $a_0 = 5$. The simulation box size was set to 24 × 20 μm in longitudinal and transverse dimension, discretized by a grid of 6,400 × 4,000 cells corresponding to a spatial resolution of Δz: 3.75 × Δ x: 5.0 nm. The nanometer foil was modeled as a fully ionized carbon plasma with density $n_e = 50\,n_c$ and 30 nm thickness (rectangular shape) using 1,000 particles per cell, which equates to a 3 nm foil at solid density.

Figure 4.1a shows the electron density and the corresponding laser field relatively early in the interaction, at ten cycles before the peak of the pulse. At this time, the electron density is clearly above critical and thus prevents the laser to penetrate through the plasma. As a result, the impinging laser is reflected and acts as a standing wave on the critical surface of the plasma. Owing to the fast oscillating $v \times B$ force of the driving field, fast electrons are generated at a frequency of 2ω and injected as a dense bunch into the plasma layer (Sect. 2.3.1). These electrons disperse in the vacuum region behind the target due to the counteracting electrostatic charge separation field built up during the interaction. These dynamics eventually result in the formation of a hot electron cloud at the target rear side linked to a huge self-induced electrostatic field, which in turn governs the ion motion over longer time scales. This scenario is characteristic for solid plasma interactions and dominates to a large extend the regime of efficient ion acceleration. However, in the ultrathin target thickness regime, the simulation indicates that the plasma turns transparent prior to the peak of the pulse, changing the interaction dynamics completely in this phase (Fig. 4.2).

Figure 4.1b shows the electron density and laser field ten cycles after the peak of the pulse has reached the target. In stark contrast to the early interaction phase, the plasma slab has turned transparent to the laser and thus interacts with a propagating rather than a standing wave. Alongside with target transparency, short, equally spaced electron bunches co-moving with the transmitted light field over long distances are evidently seen. The electrons forming these bunches are decoupled from the ion background and propagate freely in vacuum. Hence, rather than being subject to complex plasma dynamics the ejected electrons simply follow single electron motion in the transmitted electromagnetic field as discussed in Sect. 2.2.

The drastic change in plasma properties and electron dynamics becomes obvious in Fig. 4.3 showing the transmitted laser field e_x and the longitudinal electron current j_z measured two micrometer behind the target. While being overdense a rather constant electron current is observed owing to the hot electron production at the critical surface as discussed earlier. However, as the plasma turns transparent, periodically generated electron bunches formed at the laser plasma interaction region are injected into the transmitted laser field and effectively escape from the bulk plasma. Moreover, the ejected bunches can be seen to be located in the wave buckets (i.e. $e_x = 0$) of the driving laser field (Figs. 4.1 and 4.3). Note that, although initially injected "coherently", at every half cycle into the transmitted light field, each bunch

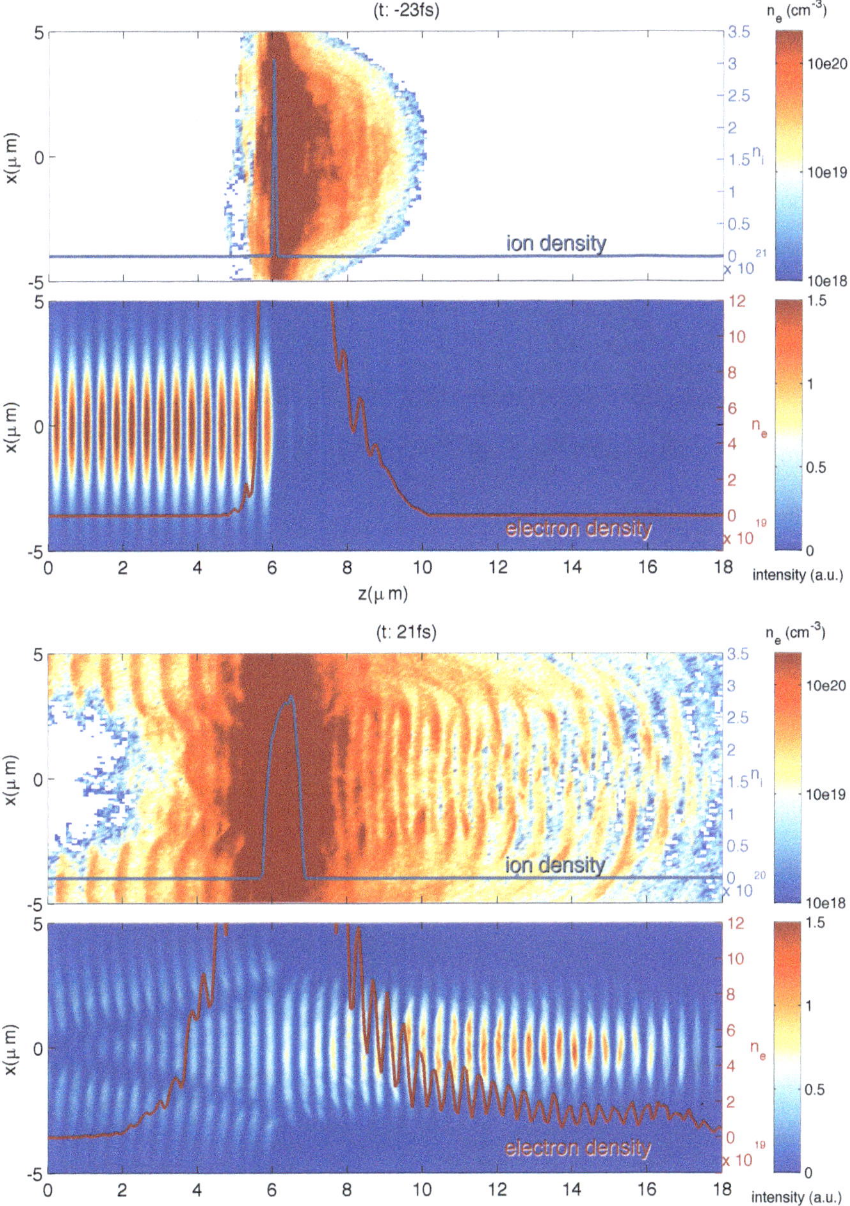

Fig. 4.1 *PIC simulation snapshots (electron density, laser field)*. The simulation shows two different phases of the interaction: (**a**) in the early phase, the plasma layer is opaque to the laser whereas at later times (**b**) the plasma turns transparent. In this phase, electrons bunches are driven out effectively

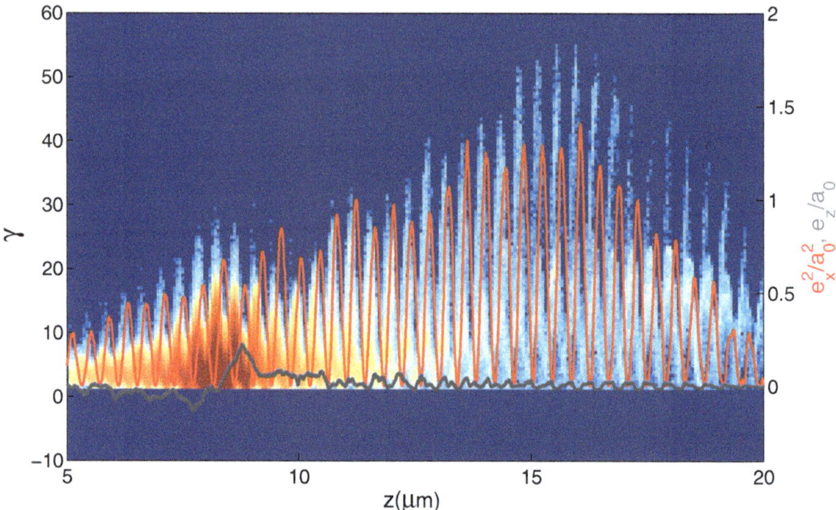

Fig. 4.2 *Electron energy space in the transparency regime*. Electrons are driven by the transmitted light field and have $\gamma \sim [1, 1 + 2a_0^2]$, consistent with what is expected from single electron dynamics

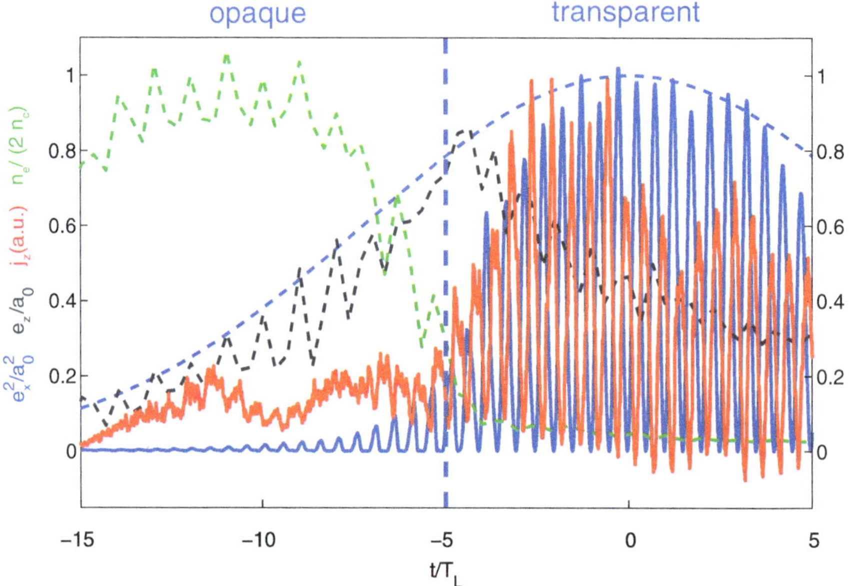

Fig. 4.3 *Temporal evolution of the laser–nanofoil interaction*. Laser intensity (e_x^2) and electron current (j_z) are measured 2 μm behind the target (averaged in transverse dimension over 4 μm, time resolution 8.8 as). The peak electron density (n_e) and the peak charge separation field (e_z) are resolved at the target position (1 μm transverse average, time resolution 0.88 fs)

consists of electrons of different energies, which in turn result in different electron trajectories in the transmitted light field. Hence, one would not expect to observe a macroscopic bunch structure after a few micrometer of propagation. Nonetheless, as clearly seen in simulation, the electron density of the ejected electrons remains bunched even after significant distances.

The macroscopic bunching of the extracted electrons can be well understood taking into account the nonlinear electron quiver motion in the transmitted electromagnetic light field. Consider a plane wave with vector potential $a_x = a_0 \sin(\tau + \phi_0)$ and $\tau = t - z$. The electric field is $e_x = a_0 \cos(\tau + \phi_0)$ with field maxima at $\tau_{max} = n\pi - \phi_0$ and zero points at $\tau_{min} = (2n + 1)\pi/2 - \phi_0$. The time-dependent phase slippage $\tau(t)$ of the electron motion is highly nonlinear and the time interval an electron spends within a given phase intervall is characterized by $dt/d\tau$. From single electron dynamics we know that $d\tau/dt = \kappa_0/\gamma$, hence using Eq. 2.16

$$dt/d\tau = \gamma/k_0 = 1 + \frac{1}{2\kappa_0^2}\left(1 - \kappa_0^2 + [a_0 \sin(\tau + \phi_0) + \alpha_0]^2\right) \qquad (4.1)$$

with parameter $\kappa_0, \alpha_0, \phi_0$ given by the initial momentum and phase of the injected electron defined in Eq. 2.17. Independent of these parameters, Eq. 4.1 reaches its maximum at the zero points of the driving field τ_{min} and its minimum at the points of maximal field τ_{max}. Considering a large number of electrons, this directly translates to density peaks located in the wave buckets of the driving field and density minima at the peaks of the driving field, consistent to what is seen in the simulation (Fig. 4.2). It is the nonlinear phase slippage that imprints to a statistical ensemble of electrons a macroscopic structure.

The observed dynamics are clearly very different from the theoretically proposed, highly idealized scenario of dense electron mirror generation from ultrathin foils (Sect. 2.3.2). Apart from the fact that in the simulated configuration the laser pulse intensity is still somewhat weak to overcome the restoring electrostatic fields, the fundamental difference stems from the (adiabatic) Gaussian rise of the laser pulse employed here. The step-like onset of the laser pulse used in the idealized, theoretical studies avoids electron heating, thus preserves the delta-like character of the nm foil, which is crucial for the formation of a coherent, nanometer thin relativistic structure. While this scheme may be accessible with upcoming few cycle high power laser systems [1, 2], we shall see in Chap. 5 that dense electron mirrors can still be created in a slightly different regime using nowadays (many cycle) high power laser pulses.

In summary, the simulation suggests strikingly different electron dynamics in the opaque and transparent interaction regime. As soon as the plasma turns transparent, electrons are effectively driven in the transmitted laser field and are accelerated to high energies, well described by single electron dynamics. Clearly, the transparent phase favors high intensities and extremely thin foils. Thus, depending on the foil thickness and the intensity of the driving laser pulse, the opaque or transparent phase prevails. The two different regimes should be clearly distinguishable in experiment as we would expect to see a drastic change in the spectral distribution of the high energetic electrons as soon as target transparency sets in.

4.2 Experimental Setup

LANL

Over the course of this Ph.D. thesis, multiple beam times were carried out at the Trident laser located at the Los Alamos National Laboratory. The Trident laser delivers ~80 J of energy in a pulse of ~500 fs FWHM duration at a central wavelength of 1,053 nm (Sect. 3.1.2). The laser pulse was focused with a f/3 off-axis parabolic mirror to a ~6 μm FWHM focal spot corresponding to an averaged (peak) intensity of $2 \times 10^{20}\,\mathrm{W/cm^2}$ ($4 \times 10^{20}\,\mathrm{W/cm^2}$).

Free-standing foils with thicknesses ranging from μm to only a few nm were used as a target. To cover such a broad range of foil thicknesses, different target materials were used: In the regime of few nm thin foils, DLC targets with density of $2.7\,\mathrm{g/cm^3}$ were employed and for thicker targets, diamond foils with density $3.5\,\mathrm{g/cm^3}$ were used.

The exact configuration of the electron and ion spectrometers varied slightly in different beam times. The experimental setup shown in Fig. 4.4 illustrates the configuration used in the beam time in April 2009, as the vast majority of the data presented in this section was measured during that campaign. In that beam time, electrons were measured using four identical magnetic spectrometers (Sect. 3.3.2), probing the electron distribution at 0° with respect to the target normal direction as well as at 8° off normal direction both along and perpendicular to the laser polarization axis. Each spectrometer was equipped with an image plate (Fujifilm BAS-TR) detector (Sect. 3.3.4), which were readout using a commercial scanner system (Fujifilm FLA-7000). Ions were measured simultaneously at 8° with respect to target normal direction using a Thomson parabola spectrometer. The electron (ion) spectrometers were fielded 1.1 m (1.3 m) away from the target resulting in acceptance angles of 4×10^{-6} sr (10^{-8} sr) for the spectrometers respectively.

MBI

The experiment was performed at the 30TW Ti:sapph laser system located at the Max Born Institute, delivering 0.7 J of energy in a pulse of 45 fs FWHM duration at

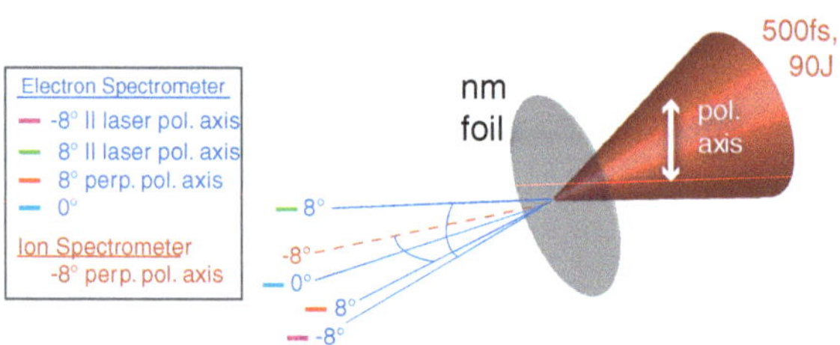

Fig. 4.4 *Experimental setup LANL* showing the experimental configuration of the Trident campaign carried out in April 2009

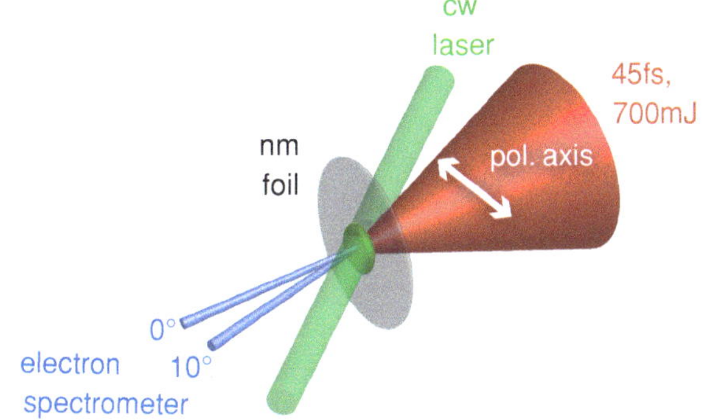

Fig. 4.5 *Experimental setup of the MBI campaign*

a central wavelength of 810 nm on target (Sect. 3.1.2). The laser pulse was focused with a f/3 off-axis parabolic mirror to a 3.5 μm FWHM focal spot corresponding to a peak intensity $5 \times 10^{19}\,\mathrm{W/cm^2}$.

Electrons were measured using two identical magnetic spectrometers, installed at 0° and 10° with respect to the target normal direction (Fig. 4.5) at a distance of 0.68 m away from the target (solid angle: $\sim 5 \times 10^{-5}$sr). An optical imaging system in combination with a scintillator screen (Biomax MS, Kodak) was used as a detector (Sect. 3.3.5).

A CW laser (Verdi, Coherent) was installed at the target chamber to be able to remove the surface contaminant layer prior to a target shot via laser target heating (Sect. 3.2).

Astra Gemini

In parallel to the backscatter experiment presented in Chap. 5, the electron distributions generated from the interaction with nanoscale targets were measured along target normal direction in the experimental campaign in 2010/2011. The experimental configuration is presented in great detail in Sect. 5.1. The Multi-Spectrometer setup was utilized to record the electron distributions and is described in Sect. 3.3.3.

4.3 Ion Measurements

It is instructive to review the key results obtained from the ion acceleration experiments, which were carried out at these laser systems in parallel to the presented work. Figure 4.6a shows the dependence of the C^{6+} ion cutoff energies on the target thickness observed at the LANL laser facility. Reducing the target thickness from the μm to sub-μm scale gives rise to a strong increase in the ion cutoff energies.

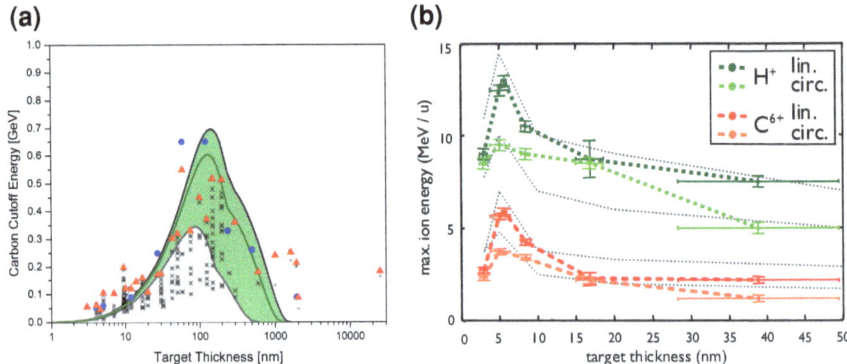

Fig. 4.6 *Ion energies observed at the LANL,* **a** *and MBI,* **b** *laser facilities*. Courtesy of M. Hegelich and A. Henig

Unprecedented high C^{6+} ion energies of up 0.5 GeV were observed at a target thickness range of 50–200 nm [3, 4]. However, reducing the target thickness further does not benefit the ion acceleration and low cutoff energies are observed from ultrathin, nm scale targets.

On the contrary, at the MBI laser system, the optimal target thickness with regard to ion acceleration is in the range of a few nm, only (Fig. 4.6b). Best results were obtained using 5 nm thin DLC foils, in which case the areal density of the target is matched to the intensity of the laser pulse [5–7]. Here, proton energies ranging up to 13 MeV and C^{6+} energies up to 71 MeV were achieved [8, 9]. Reducing the target thickness further, again, results in a clear reduction in ion energies.

At Astra Gemini, the target thickness dependence is not that clear and still is under current investigations. Proton energies on the order of ~20 MeV and carbon C^{6+} energies around ~100 MeV were typically observed from nanoscale foils. These values are clearly well below those expected from a petawatt class laser system and thus achieving higher energies is subject to ongoing research.

4.4 Target Thickness Scan

More than just presenting a few selected electron spectra, we shall discuss the overall dependencies observed from the target thickness scans. To characterize the recorded electron distributions and allow for a comparison among a huge number of different electron spectra, we deduce the electron mean energy $\langle E \rangle$ from each spectrum within the resolved spectral range:

$$\langle E \rangle = \frac{\int_{E_{min}}^{E_{max}} E \frac{dN}{dE} dE}{\int_{E_{min}}^{E_{max}} \frac{dN}{dE} dE} \tag{4.2}$$

where E_{min}, E_{max} denote the minimal (maximal) electron energy measured by the spectrometer. In the case of an exponential distribution $dN/dE = N_0 \exp(-E/T_{hot})$, we can derive an analytical expression:

$$\langle E \rangle = \frac{T_{hot} + E_{min} - (T_{hot} + E_{max})e^{-(E_{max}-E_{min})/T_{hot}}}{1 - e^{-(E_{max}-E_{min})/T_{hot}}} \quad (4.3)$$

Thus, if the resolved spectral range $\Delta E = E_{max} - E_{min}$ is large with respect to the hot electron temperature T_{hot}:

$$\langle E \rangle \approx T_{hot} + E_{min} \quad (4.4)$$

This method provides means to characterize even those spectra, which do not follow an exponential distribution. On the other hand, in the case of an exponential electron distribution, the hot electron temperature can be trivially deduced from $T_{hot} = \langle E \rangle - E_{min}$, which allows for a direct comparison to the theoretical literature.

4.4.1 Experimental Observations

The electron spectra measured from μm to nm scale targets usually follow exponential distributions with energies typically ranging up to 1–5 MeV (MBI), 10–20 MeV (RAL) and 30–50 MeV (LANL). A few characteristic spectra are shown in Fig. 4.7. Yet, a more nuanced picture on the target thickness dependence can be obtained when comparing the electron mean energies of the measured spectral distribution. Figure 4.8 summarizes the electron measurements carried out at the LANL, MBI and RAL laser system depicting the electron mean energies deduced from more than two hundred electron spectra. The analysis reveals a strong increase in the electron mean energy as the target thickness is reduced to the nanometer scale. This enhancement in the measured electron energies is consistently observed at all three laser systems. Analogously, the ion energies can be increased considerably as the target thickness is reduced. However, this holds true up to a certain thickness optimal for ion acceleration, which strongly depends on the parameters of the driving laser pulse. Reducing the target thickness even further, clearly beyond the optimal target thickness range, the ion energies drop down considerably, while a significant increase in the electron mean energies is observed. This transition was clearly seen at the LANL and MBI laser system. At the Astra laser, however, the thickness dependence on the ion acceleration is somewhat flattened out and only a slight drop in ion energies is seen even in the case of a 5 nm foil. Likewise, although the electron mean energy increases as the target thickness is reduced, the transition is not as sharp and the increase in energy is not as high as in the other two cases.

At the LANL and MBI laser system, the target thickness could be reduced clearly beyond the optimum for ion acceleration, down to a thickness range where the ion signal breaks down completely. In this regime, a transition in the electron distributions from monotonically decaying to quasi-monoenergetic is observed (Sect. 4.5).

Evidently seen in Fig. 4.8, the spread in the recorded data increases significantly in the regime of nm scale targets, which can be understood immediately taking into account the fact that the interaction becomes increasingly more sensitive to the exact laser pulse parameters as well as target properties when reducing the target thickness. Recently, single shot FROG measurements revealed variations in laser pulse shape of the Trident laser, which could strongly affect the electron dynamics during the interaction [10]. Considerable efforts have been carried out to improve the stability of the system and to monitor the contrast and shape of the incident laser pulse by implementing a single shot autocorrelator and FROG into the system. These diagnostics were not available at the time of the experimental campaign making the interpretation of the data even more challenging.

Compared to the LANL measurements, the electron spectra obtained from the MBI and Astra laser exhibit rather high reproducibility owing to the (one order of magnitude) shorter pulse duration of the laser system in combination with a better contrast ratio on the ps time scale.

4.4.2 Theoretical Discussion

Despite the strong differences in the utilized laser systems, the presented electron measurements reveal a similar dependence on the target thickness, which shall be discussed in the following.

In the case of a thick, truly overdense target, high energetic electrons are generated at the front-side laser plasma boundary throughout the whole interaction. In this regime, it is the scale length (gradient) of the laser plasma interface rather than the thickness of the plasma that determines the dynamics of the high energetic electrons resolved by the spectrometer. Thus, we do not expect a strong dependence on target thickness and indeed that is what is observed for all laser systems over a broad range of thicknesses. The mean electron energies observed in this target thickness regime crucially depend on the laser pulse intensity as well as on the front side plasma gradient. The corresponding hot electron temperatures predicted by the scaling law recently published by [11], (Sect. 2.3.1, Eq. 2.29) almost perfectly match to the MBI and RAL measurements, while at LANL, the experimentally deduced hot electron temperature ($T^{LANL} \sim 5-8\,\text{MeV}$) considerably deviates from the theoretically expected value ($T^{LANL}_{Kluge} = 1.7\,\text{MeV}$). However, this apparent mismatch can be easily understood considering the fact that this scaling law assumes a perfectly sharp laser plasma boundary. Thus, it seems just consistent with the theoretical expectation that both short pulse laser systems using a double plasma mirror for further contrast enhancement satisfy the underlying assumption of a steep interface (and hence follow the scaling), whereas the long pulse laser having considerably lower contrast on the few ps time scale (Sect. 3.1.2, Fig. 3.3) does not. On the contrary, the ponderomotive scaling by Wilks [12], (Sect. 2.3.1, Eq. 2.28) is in good agreement with the LANL measurements, whereas this scaling drastically fails for the high contrast, short pulse interactions ($T^{MBI}_{Wilks} = 1.4\,\text{MeV}$, $T^{Astra}_{Wilks} = 5.7\,\text{MeV}$).

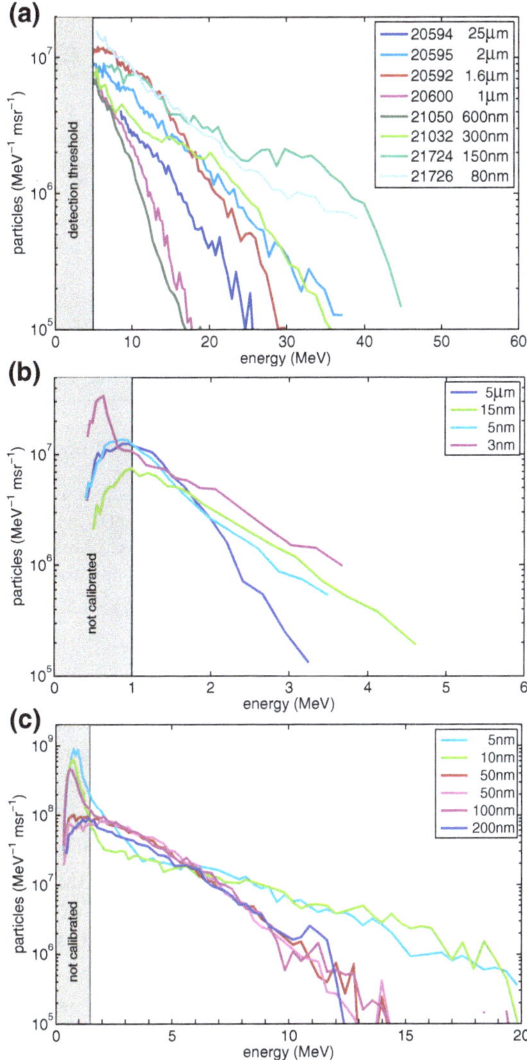

Fig. 4.7 *Typical electron spectra observed at* **a** *LANL,* **b** *MBI and* **c** *RAL*

The interaction dynamics discussed above dominate as long as $n_e > \gamma n_c$, i.e. in the case of an opaque plasma. However, by reducing the target thickness to the nm scale, plasma transparency starts to become important. The onset of transparency is determined by the time evolution of the electron density during the interaction. The exact dynamics are very complex. Insufficient contrast of the laser pulse can cause target pre-expansion and density reduction of the target prior to the interaction with the main pulse. During the interaction with the peak of the pulse, the front side density gets compressed owing to the ponderomotive force of the laser pulse

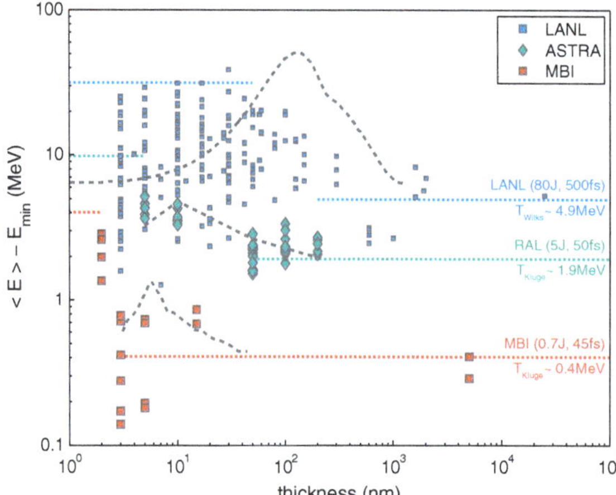

Fig. 4.8 *Electron mean energies observed from different laser systems and foil thicknesses*. The mean energies of the measured spectral distributions are renormalized to the low detection threshold of the spectrometers E_{min} (LANL: 10 MeV, MBI: 1.5 MeV, ASTRA: 2 MeV). Each of the depicted energy values $\langle E\rangle - E_{min}$ corresponds to the mean energy of an exponential spectrum with spectral slope T_{hot}. Gray, dashed lines: thickness dependence of the carbon C^{6+} energies observed at the laser systems

[13]. After a phase of compression, the density eventually drops rapidly (Fig. 4.3). Depending on the intensity of the incident pulse, the plasma turns transparent at a density well above the stationary critical density owing to relativistically induced transparency. At present, theoretical studies modeling target transparency are highly idealized (delta-like foil models [14, 15], plasma expansion models [16, 17]) and do not grasp the complex dynamics of the electron density during the interaction. Moreover, even PIC simulations have large uncertainty due to the unknown initial density profile. Nonetheless, it is clear that transparency should become increasingly more important as the thickness of the target is reduced.

As soon as transparency sets in, the incident laser pulse penetrates into the plasma and effectively couples to all electrons within the interaction volume rather than acting as a standing wave on the critical plasma surface, only. Clearly, this scenario is different from the interaction of the laser with a sharp laser plasma interface and hence the scalings laws discussed above are no longer valid. Instead, we observe a gradually rising electron mean energy along with increasing laser transmission.

An upper limit for the electron mean energy expected in the transparency regime can be derived from single electron dynamics. In this rather simplistic scenario, the final energy gain of an electron is determined by the initial phase the electron is born into the field $\gamma = 1 + a(\phi)^2/2$ (Sect. 2.24, Eq. 2.20). Assuming that electrons are continuously injected into the laser field, the mean energy is determined by the average over all phases thus $T_{Trans} = (2\pi)^{-1} \int_0^{2\pi} \gamma(\phi)\, d\phi$, which yields $T_{Trans} = 1 + a_0^2/4$ for a plane wave [18].

Now, considering the fact that the initially opaque plasma turns transparent at some time during the interaction, it is clear that the field driving the transparent target strongly depends on the transmission function of the plasma and actually may not even reach the peak field of the incident laser. While the optical shuttering properties of the expanding plasma are unknown from the experiment, the measured time-integrated plasma transmission allows making a rough estimate. Assuming a step-like transition from opaque to transparent, it is clear that the peak field of the incident laser pulse is reached in the transparent phase, only if the recorded transmission through the target exceeds 50 %. In fact, as the transition is not expected to be as sharp, even higher values may be needed.

The effect of plasma transparency is most evidently seen in the LANL data. In the target thickness range of effective ion acceleration, that is in the regime where relativistically induced transparency is expected to become important [16], we observe a gradually rising electron mean energy.[1] As the target thickness is reduced to the few nm scale, the recorded ion energies break down, while the electron mean energies gradually increase and eventually saturate at around 30 MeV. In this regime, simulations and experiments indicate that the plasma turns transparent long before the peak of the pulse [19, 20], and thus we would expect $T_{Trans}^{LANL} = 29\,\text{MeV}$. Likewise, at the MBI laser, a transmission as high as 60 % was observed from 3 nm pre-heated targets [8]. Thus, we argue that the target turns transparent before the peak of the pulse and accordingly we observe good agreement with the free electron limit ($T_{Trans}^{MBI} = 3.7\,\text{MeV}$). The electron mean energies observed at the Astra Gemini laser, however, do not reach as high energy values as one would expect from the high peak intensity of the laser pulse. While the reason is not obvious, the observation is still consistent with the measured rather low transmission of 25 through a 5 nm thin foil.[2] Hence, at Astra Gemini, it seems that transparency is not expected to play a dominant role even for the thinnest targets. In the case of a 5 nm foil, the reported transmission value allows for peak fields not much higher than $a_0/2$ in the transparent phase, hence electron mean energies of $T_{Trans}^{Astra} = 10\,\text{MeV}$ would be expected.

The slight mismatch with the observed electron energies indicates that the simple free electron scaling is only valid in the fully transparent regime.

In summary, the observed increase in the electron energies can be well explained by the onset of plasma transparency. It is worth noting that this interpretation is supported by the fact that the measured ion energies decrease considerably as the target thickness is reduced to only a few nm. This observation can be intuitively understood considering the fact that the PIC simulation (Fig. 4.3) indicates a clear drop in the electrostatic field upon transparency.[3] As a reduction in target thickness causes the plasma layers to turn transparent increasingly early, a considerable decrease of the ion energies would be expected in excellent agreement with the experimental obser-

---

[1] While in this thickness regime only little transmission values were reported, this may be very well explained by enhanced laser absorption (and hence effective ion acceleration).

[2] Private communication, W. Ma.

[3] A simple way to explain this behavior is that, upon transparency, the ponderomotive force (radiation pressure) of the laser on the plasma layer vanishes ($f_p \propto \nabla E_A^2$).

vations. On the contrary, the collapse of the counteracting, longitudinal field allows a major fraction of the electrons to escape from the target more effectively. In that sense, both electron and ion measurements are complementary and are in line with the theoretical interpretation.[4]

4.5 Electron Blowout

While the electron spectra observed from thick targets follow a monotonically decaying exponential distribution, a clear departure from this purely thermal spectral behavior is observed in the extreme thickness regime of a few nm thin foils. Here, the spectral shape clearly changes from exponential to quasi-monoenergetic. The transition to the blowout regime was consistently observed at the LANL and MBI laser facility and will be presented in detail in following section.

4.5.1 LANL

Figure 4.9 shows the electron spectra measured in target normal direction from 5 to 3 nm thin foils. In contrast to the typically observed exponential electron signal, the recorded electron spectra clearly follow a quasi-monoenergetic spectral shape. Simultaneously, as mentioned before, the ion signal drops down drastically, to below

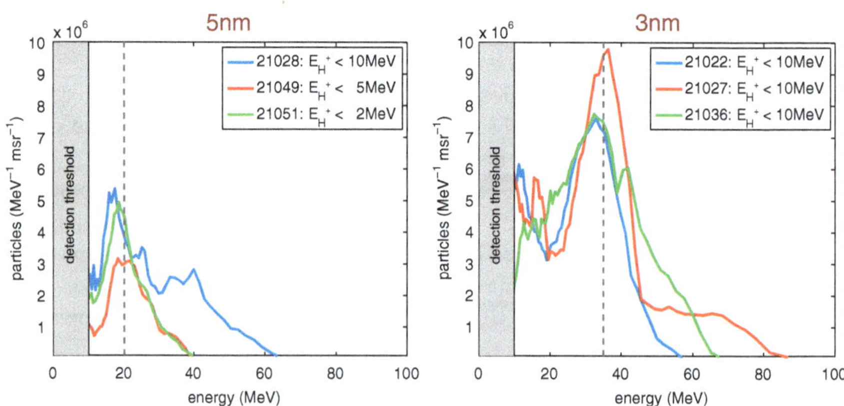

Fig. 4.9 *LANL electron blowout spectra measured at 0° from 5 to 3 nm DLC foils.* No proton signal was measured from all of these shots. The low energy detection threshold of the Thomson parabola setup is given in the figure label

[4] The counteracting, electrostatic fields in the semi-transparent regime may also explain the mismatch with the free electron scaling found at the Astra Gemini laser.

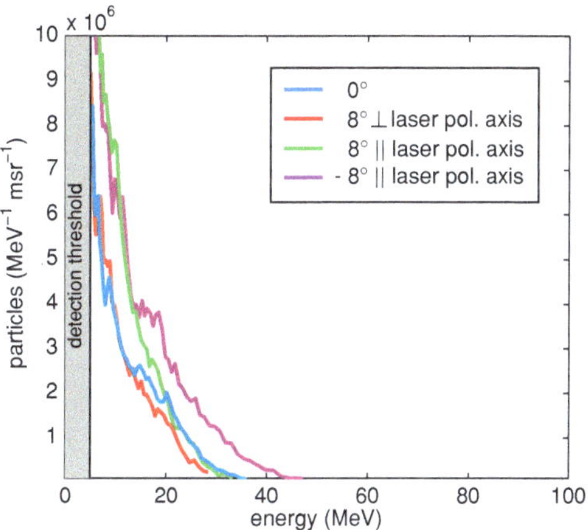

Fig. 4.10 *LANL electron spectra observed from a 300 nm target measured simultaneously in different directions*

the low energy detection threshold in this extreme target thickness regime. While in the case of a 5 nm thin foil, the observed electron distributions are peaked at ~20 MeV, the monoenergetic feature evolves to ~35 MeV when irradiating a 3 nm foil. Moreover, electrons are extracted from the 3 nm thin foil more efficiently, as in this case, the number of electrons recorded within the monoenergetic feature is significantly increased. Overall, this spectral behavior was observed multiple times from various target shots, demonstrating a remarkable reproducibility. To get an insight into the spatial distribution of the measured electron signal, four identical electron spectrometer were installed at different angles as depicted in Fig. 4.4. As expected from a purely thermal distribution, the electron spectra measured from a 300 nm thick foil (Fig. 4.10) are almost identical in all directions and in particular independent of the laser polarization axis. Moreover, the signal measured on-axis that is in target normal direction does not exhibit any characteristics different from the off-axis signal and is even identical in magnitude. This observation is of particular interest for ion acceleration studies in the thick target regime, as it is sufficient to probe the electron distribution at some off-axis angle for those target shots, a configuration which allows measuring the ion distribution in target normal direction synchronously. In contrast to the isotropic distributions observed from μm scale targets, the electron distributions obtained from ultrathin foils are nonuniform. Figure 4.11 shows the electron spectra obtained from the 3 to 5 nm target shots measured at different angles simultaneously. While the exact signal does substantially vary in off-axis direction, highest electron energies are observed in target normal direction.

In addition to the spectral measurement obtained from the multiple electron spectrometer setup, an image plate stack detector was used to record a footprint of the electron beam generated from a 3 nm thin foil. The detector consists of twelve image plate films each separated by a stopping layer made of aluminum of varying thickness (1–3 mm). The assembled stack was positioned ~5 cm behind the target. Owing to the continuous stopping characteristic of electrons in matter, the spectral deconvolution of the data requires sophisticated MonteCarlo simulations [21] and shall not

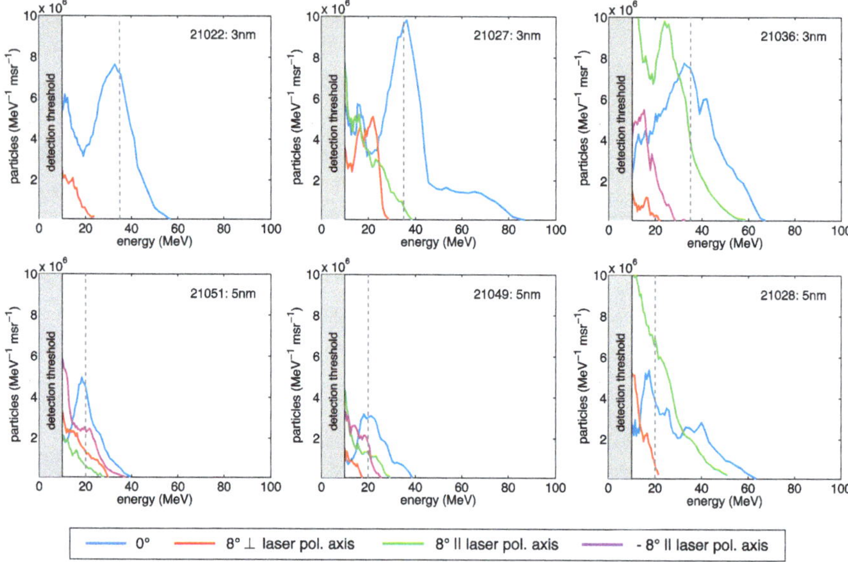

Fig. 4.11 *LANL electron blowout spectra: Angular dependence.* Electron spectra measured simultaneously at 0 and 8° with respect to the laser axis and in different directions with respect to the laser polarization axis

be discussed here. Instead, the raw data of a single image plate positioned behind 44 mm thick alluminum is presented in Fig. 4.12. The data suggests that the electron signal is predominantly directed forward and is enhanced along the laser polarization direction, which is in good agreement with the multiple spectrometer measurements. The experimental findings allow making an estimate on the electron beam characteristics. Taking the average of the 3 nm shots presented in Fig. 4.9, the peak energy is $E_{peak} = (33.9 \pm 1.2)$ MeV and the energy spread (FWHM value) $\Delta E_{FWHM} = (23.5 \pm 4.1)$ MeV. Assuming an emission cone with half apex angle of 5°, the charge within the FWHM energy spread of the measured electron beams is $Q = (542 \pm 70)$ pC.

4.5.2 MBI

Consistent with the observations from the LANL experiment, electron distributions of apparently different, non-exponential shape were measured at the MBI laser system when irradiating nanometer foils with ever decreasing thickness (Fig. 4.13). In the thickness regime of efficient ion acceleration, exponentially decaying electron distributions were found. The recorded spectra were (within typical fluctuations) identical and independent of the observation angle. Even for the thinnest free-standing DLC foils available (3 nm), the electron spectra did not reveal any significant change

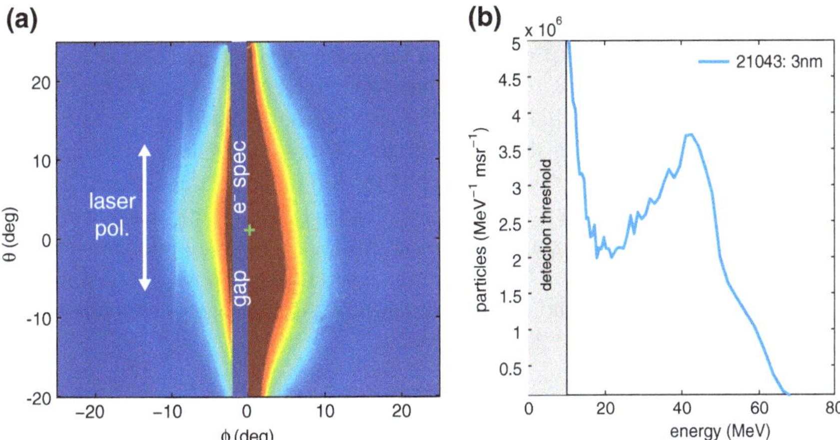

Fig. 4.12 *LANL electron blowout spectra: Footprint.* The depicted electron beam profile was recorded behind 44 mm thick aluminum. Hard X-rays, which could potentially penetrate the stopping material and therefore cause a misleading signal, can be neglected due to the low image plate sensitivity for photon energies above few tens of keV. The electron distribution recorded simultaneously along the target normal direction using a magnetic spectrometer is shown in (**b**)

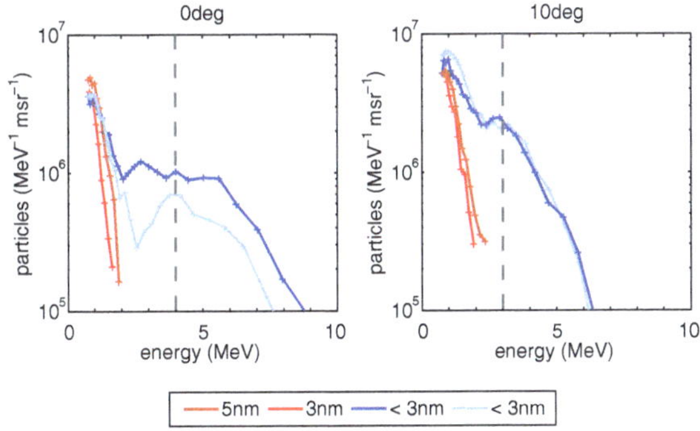

Fig. 4.13 *Electron blowout spectra: MBI.* The target thickness of the 3 nm DLC foils is reduced using target pre-heated. The exact target thickness is unknow. In the following, we refer to those target shots as 2 nm +/− 1 nm thin targets

whereas moderate ion energies were still achieved. To reduce the target thickness even further, 3 nm thin foils were heated in the target chamber using a CW laser in order to remove the hydrogen contaminant layer from the target surface prior to the laser shot. Despite the thermal stability of the DLC material, the controlled heating of such an extremely thin free-standing foil is challenging and was carried out with great care. In order to find appropriate heating parameters, the CW laser power and the irradiation (heating) time was increased systematically in subsequent laser target shots. Heating the foil with 200 mW output power for 30–50 s (FWHM focal spot size ~200 μm), a drastic change in the electron distribution is observed. Here, an additional spectral component clearly above the thermal electron background is found in the distributions, peaked at ∼4 MeV in target normal direction and at a slightly reduced energy of ∼3 MeV at 10°. Moreover, when increasing the target heating further, we recover the exponentially shaped, low temperature distributions as observed from regular target shots. In this case, the target imaging system monitoring the heating process displayed the burn through of the foil in the central region (Sect. 3.2, Fig. 3.6) and therefore the measured, residual electron signal originates from the low intensity side wings of the interaction region. Although the exact thickness remains unknown in the case of target heating, ion measurements denote significantly less proton signal and thereby indicate that upon heating, hydrogen contaminants are effectively removed from the target surface.

4.5.3 Theoretical Discussion

The simulation presented in Sect. 4.1 clearly indicates the formation of energetic electron bunches, accelerated in the transmitted laser field, which is in reasonable agreement with the observation of target transparency and enhanced electron signal from ultrathin foils. However, to extract the final energy distribution of the generated electron bunch train and compare the simulation with the electron spectra observed in the experiments considerably more computational effort is needed. The complexity stems from the fact that in the transparency regime, large simulation box sizes are required as the accelerated electrons co-propagate with the driving laser field over long distances. In order to determine the final energy gain of the electrons the laser pulse needs to fully slip over the relativistically moving electrons which translates to hundreds of μm to even mm long distances and thus is very challenging given the high resolution needed to resolve the nm foil at the beginning of the interaction.

Such a full scale simulation was recently carried out by [22] modeling the interaction of the LANL laser with a few nm thin foil using simulation parameters close to the experimental configuration reported in [23]. Making use of advanced computational techniques such as adaptive mesh refinement and a moving window, the simulation was run until the laser had fully overtaken all electrons.

Glazyrin et al. [22] report that in fact to explain the observed quasi-monoenergetic electron distributions, ionization dynamics have to be taken into account. A direct comparison of the energy spectra obtained from a fully pre-ionized 5 nm thin plasma target (typically used in PIC simulations) and an initially neutral carbon foil is shown in Fig. 4.14. The spectral peak observed from the initially neutral foil is remarkably close the observed quasi-monoenergetic feature while the full plasma simulation does not reveal a secondary high energetic spectral peak. Moreover, the peaked spectral component could not be observed from a rather thick 42 nm target in agreement with the experimental observation.

In the simulations, field ionization is accounted for dynamically using the Ammosov-Delone-Krainov (ADK) model. As a result, electrons from different atomic shells are subsequently released and injected into the laser field as the pulse

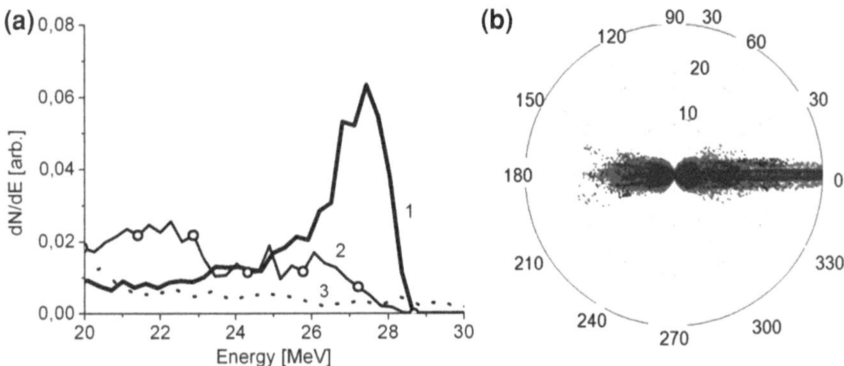

Fig. 4.14 *Electron blowout spectrum: PIC simulation spectra and emission characteristic.* **a** Electron distribution recorded after the pulse has fully slipped over the high energetic electrons and corresponding angular distribution **b** Both figures are taken from [22]

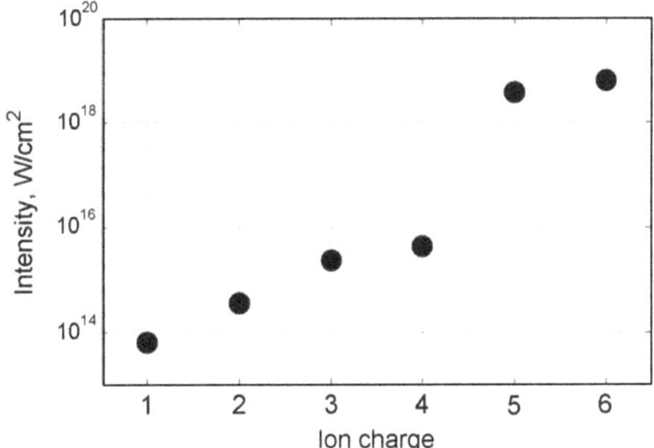

Fig. 4.15 *Field ionization of carbon* [22]

rises in intensity. The dynamic change in the electron population is observed to give rise to essentially two different groups of electrons. One species is formed by electrons originating from outer atomic shells, which are born early in the interaction, at sub-relativistic intensities due to their low ionization potential (Fig. 4.15). These electrons build up a plasma, interacting with the laser field not considerably different from what is seen in the case of an initially fully ionized plasma target. A second group appears from the inner shell electrons, which is released late, at relativistic intensities, close to the peak of the pulse. Upon the interaction with the peak laser field, these electrons still have a narrow spread in phase space, as they have not gone through many oscillation cycles during the rise of the pulse and thus can be accelerated in a narrow spectrum.

While the simulation shows good agreement with the experimental results, the theoretical description of the acceleration process is still somewhat unclear. From the

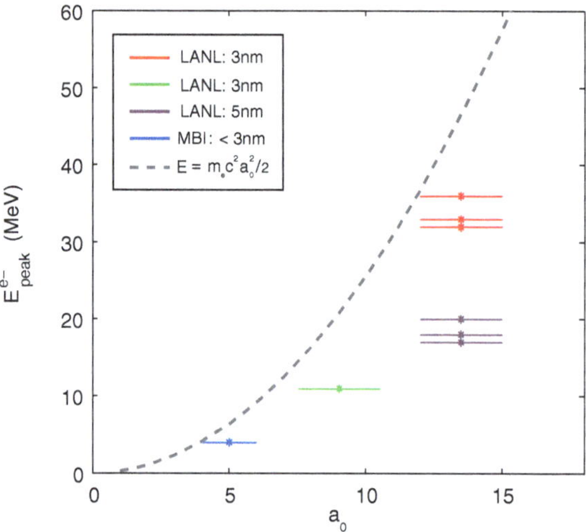

Fig. 4.16 *Electron blowout energy scaling*. Electron blowout peak energies observed from different laser–nanofoil configurations

discussion given in Sect. 2.2.4, it is clear that considerable energy gain can be achieved from the abrupt (nonadiabatic) seeding of electrons into the propagating laser field. Field ionization takes place on short time scales compared to the laser period and thus provides such a mechanism. In fact, theoretical studies show that electrons born from high Z atoms into the high intensity region of a strong laser pulse can be accelerated to GeV energies directly in the laser field [24, 25]. However, the ionization potentials of carbon atoms are comparably low and thus the energies expected from the equivalent process ($\sim$1 MeV [22]) would not explain the experimental observation.

The rapid formation of high energetic electron bunches from the plasma background, nonetheless, may act analogous to the ionization event. Owing to the collective plasma fields built up during the interaction, the electron plasma stays bound to the ion background and only a small fraction of electrons is released at every half cycle as a bunch. These electrons rapidly acquire high momentum during the bunch formation process, which allows them to overcome the counteracting charge separation field. The dynamics of those electrons are observed in simulation to change rapidly from stochastic plasma motion to the single electron dynamics and thus may very well undergo a non-adiabatic seeding similar to the ionization event. In fact, it was pointed out by [26] that the break up of adiabaticity is a key feature of the electron bunch formation in an overdense plasma. Following this line of thought, the abrupt injection of an electron bunch from the plasma into the peak of the propagating laser field would result in a final energy gain $E_f = m_e c^2 a_0^2/2$. This quadratic scaling even holds when considering a focused laser pulse [27, 28].

Figure 4.16 summarizes the experimentally observed electron peak energies measured from different laser nanofoil configurations at the MBI and LANL laser. The experimental data clearly indicates a quadratic dependence on the a_0 parameter and is strikingly close to the single electron scaling motivated above. However, in a plasma, counteracting electrostatic fields originating from the ion background are built up during the interaction and thus we observe slightly reduced electron energies. Irradiating a thicker 5 nm foil, these fields slightly increase and therefore lower energy values are found. This considerable drop in energy clearly illustrates that the efficient electron blowout requires ultrahigh intensities combined with ultrathin foils. For a MBI type laser, the use of sub 3 nm foils is indeed crucial.

While the measured electron energies can be remarkably well explained by the $a_0^2/2$ scaling, the underlying single electron model does not fully grasp the complexity of the interaction. In fact, in the single electron model given above, the energy gain of the electron is directly bound to a transverse momentum gain ($p_\perp \sim a$) and thus we would expect the formation of two electron beams with emission angle $\tan\theta \sim 2/a_0$ from the interaction, which for LANL (MBI) corresponds to 8° (22°). More sophisticated models (using higher order field components [29, 30]) yield different ejection angles, however, none of them would explain the experimentally observed narrow beam emission in *forward* direction (highest energies were consistently observed in 0° in both experiments). Standard PIC simulations indeed indicate the off-axis emission of two electron beams as a result of the periodic generation of relativistic electron bunches in alternating transverse directions [31]. This characteristic emission pattern is also seen at early times in the large scale simulation by [22]. However, it is reported that this angular dependences blurs out after long propagation distances due to space charge effects and thus eventually, a single beam in forward direction is observed.

To address this question in detail and resolve the angular dependence in future experiments more accurately, a novel, wide angle electron spectrometer was developed in the framework of this thesis, capable of resolving electron energies within a detection angle of ~25° in a single shot. Preliminary experiments, however, did not exhibit the off-axis emission of collimated electron beams, which hints that the off-axis emission pattern may be indeed lost after long propagation distances.

4.5.4 Competing Mechanisms

The observation of quasi-monoenergetic electron beams from laser nanofoil interactions is an absolutely new discovery. While we find strong indication that these electrons are accelerated directly by the laser pulse, we shall critically consider alternative interpretations, for example the laser wakefield acceleration (LWFA) mechanism in an underdense plasma could also explain the experimental results. In fact, several groups have investigated the generation of collimated electron jets from solid density targets by making use of low laser pulse contrast conditions (or deliberately introducing a pre-pulse) to create a short, low density plasma from a solid target.

For instance, in normal incidence configuration, quasi-monoenergetic electron beams of rather low energy (~0.6 MeV) were observed from a pre-exploded foil target (7.5 μm), which was ionized by an intense pre-pulse ~0.5 ns prior to the arrival of the main pulse [32]. Likewise, a low divergent electron beam was observed at the LANL laser using low ASE contrast (no pulse cleaning, target ionization at ~0.5 ns) and a 100 nm thick target irradiated at oblique incidence angle [33]. However, in the latter experiment, the spectral distribution of the generated electron beam was poorly resolved (and from the given data points seems exponential). In all experimental studies investigating the electron acceleration from pre-exploded foil targets [32–34], plasma density measurements or just estimates on the plasma pre-expansion suggest the interaction of the main pulse with an heavily expanded plasma of below critical density and few 100 s μm scale length. Hence, it was argued that in such an interaction scenario, the main pulse drives a wakefield in the expanded, underdense plasma, accelerating electrons to MeV energies in a low divergent beam (laser wakefield acceleration).

However, in clear contrast to those studies mentioned above, the experimental results obtained in this thesis were performed with ultrahigh contrast laser pulses and nm scale targets. Yet, the pre-expansion of the irradiated nm foils in advance of the main pulse is essentially unknown. To get an idea, the contrast curves of the laser systems (Sect. 3.1.2, Fig. 3.3) can be used as a guide to estimate the onset of the plasma formation. From those curves we deduce that in the case of the MBI laser pulse ionization should not take place earlier than -2 ps prior to the peak whereas at the LANL experiment the target may already ionize at ~ -50 ps in advance of the main pulse. Following the discussion given in [33], we would expect a 3 nm thin (470 n_c) target to expand to 30 nm (44 n_c). This estimate is consistent with the density scaling inferred from high harmonic measurements from nm foils using the same laser system [35]. Hence, for the MBI experiment, we have strong indication that even a few nm thin foils are truly overdense at the arrival of the main pulse and thus any LWFA scenario does not apply. In the case of the LANL experiment, we estimate an expansion of ~4 μm (hence $n_e \sim n_c$) for an initially 3 nm thin foil and thus cannot exclude the interaction with an underdense plasma from those simple estimates. However, even in such a situation, the generation of a quasi-monoenergetic electron distribution can still not reasonably be explained by a LWFA scenario. In fact, experiments at a very similar laser system (Vulcan laser: 160 J, 600 fs, $a_0 \sim 15$) using gas jets covering a wide span of densities ($5 \times 10^{18}\,\mathrm{cm}^{-3} - 1 \times 10^{20}\,\mathrm{cm}^{-3}$) displayed—without exception—monotonically decaying electron distributions [36]. This holds true when using foam targets of even higher, close to critical densities ($0.9 - 3\,n_c$) [37]. Another, completely different process relevant in this regime is the "direct laser acceleration" (DLA) [38], which under the right conditions can prevail over the LWFA mechanism. Still, this process does not give rise to a quasi-monoenergetic electron distribution [36, 39].

To date, no simulation or even experimental study exists that would show the generation of a quasi-monoenergetic electron beam from an underdense plasma using a 500 fs long laser pulse. Taken all together, we conclude that a LWFA scenario does not match to our interaction parameters and cannot explain the experimental observations.

References

1. Daniel H, Laszlo V, Raphael T, Franz T, Karl S, Vladimir P, Ferenc K (2009) Generation of sub-three-cycle, 16 tw light pulses by using noncollinear optical parametric chirped-pulse amplification. Opt Lett 34(16):2459–2461
2. Major Z, Trushin S, Ahmad I, Siebold M, Wandt C, Klingebiel S, Wang TJ, Fülöp JA, Henig A, Kruber S, Weingartner R, Popp A, Osterhoff J, örlein RH, Hein J, Pervak V, Apolonski A, Krausz F, Karsch S (2009) Basic concepts and current status of the petawatt field synthesizer-a new approach to ultrahigh field generation. Rev Laser Eng 37(6):431–436
3. Henig A (2010) Advanced approaches to high intensity laser-driven ion acceleration. Ph.D. thesis, Ludwig-Maximilians-Universität München (LMU)
4. Jung D (2012) Ion acceleration from relativistic laser nano-target interaction. Ph.D. thesis, Ludwig-Maximilians-Universität München (LMU)
5. Esirkepov T, Yamagiwa M, Tajima T (2006) Laser ion-acceleration scaling laws seen in multiparametric particle-in-cell simulations. Phys Rev Lett 96:105001
6. Yan XQ, Lin C, Sheng ZM, Guo ZY, Liu BC, Lu YR, Fang JX, Chen JE (2008) Generating high-current monoenergetic proton beams by a circularlypolarized laser pulse in the phase-stableacceleration regime. Phys Rev Lett 100:135003
7. Rykovanov SG, Schreiber J, Meyer ter Vehn J, Bellei C, Henig A, Wu HC, Geissler M (2008) Ion acceleration with ultra-thin foils using elliptically polarized laser pulses. New J Phys 10(11):113005
8. Steinke S, Henig A, Schnürer M, Sokollik T, Nickles PV, Jung D, Kiefer D, Hörlein R, Schreiber J, Tajima T, Yan XQ, Hegelich M, Meyer ter V, Sandner W, Habs D (2010) Efficient ion acceleration by collective laser-driven electron dynamics with ultra-thin foil targets. Laser Part Beams 28(01):215–221
9. Henig A, Steinke S, Schnürer M, Sokollik T, Hörlein R, Kiefer D, Jung D, Schreiber J, Hegelich BM, Yan XQ, Meyer-ter Vehn M, Tajima T, Nickles PV, Sandner W, Habs D (2009) Radiation-pressure acceleration of ion beams driven by circularly polarized laser pulses. Phys Rev Lett 103(24):245003
10. Palaniyappan S, Shah RC, Johnson R, Shimada T, Gautier DC, Letzring S, Jung D, Horlein R, Offermann DT, Fernandez JC, Hegelich BM (2010) Pulse shape measurements using single shot-frequency resolved optical gating for high energy (80 j) short pulse (600 fs) laser. Rev Sci Instr 81(10):10E103
11. Kluge T, Cowan T, Debus A, Schramm U, Zeil K, Bussmann M (2011) Electron temperature scaling in laser interaction with solids. Phys Rev Lett 107:205003
12. Wilks SC, Kruer WL, Tabak M, Langdon AB (1992) Absorption of ultra-intense laser pulses. Phys Rev Lett 69(9):1383–1386
13. Kemp AJ, Sentoku Y, Tabak M (2008) Hot-electron energy coupling in ultraintense laser-matter interaction. Phys Rev Lett 101(7):075004
14. Vshivkov VA, Naumova NM, Pegoraro F, Bulanov SV (1998) Nonlinear electrodynamics of the interaction of ultra-intense laser pulses with a thin foil. Phys Plasmas 5(7):2727–2741
15. Andrea M, Silvia V, Tatyana VL, Francesco P (2010) Radiation pressure acceleration of ultra-thin foils. New J Phys 12(4):045013+
16. Yin L, Albright BJ, Jung D, Shah RC, Palaniyappan S, Bowers KJ, Henig A, Fernndez JC, Hegelich BM (2011) Break-out afterburner ion acceleration in the longer laser pulse length regime. Phys Plasmas 18(6):063103
17. Yan XQ, Tajima T, Hegelich M, Yin L, Habs D (2010) Theory of laser ion acceleration from a foil target of nanometer thickness. Appl Phys B 98:711–721
18. Kluge T, Gaillard SA, Flippo KA, Burris-Mog T, Enghardt W, Gall B, Geissel M, Helm A, Kraft SD, Lockard T, Metzkes J, Offermann DT, Schollmeier M, Schramm U, Zeil K, Bussmann M, Cowan TE (2012) High proton energies from cone targets: electron acceleration mechanisms. New J Phys 14(2):023038

19. Henig A, Kiefer D, Markey K, Gautier DC, Flippo KA, Letzring S, Johnson RP, Shimada T, Yin L, Albright BJ, Bowers KJ, Fernández JC, Rykovanov SG, Wu HC, Zepf M, Jung D, Liechtenstein VK, Schreiber J, Habs D, Hegelich BM (2009) Enhanced laser-driven ion acceleration in the relativistic transparency regime. Phys Rev Lett 103(4):045002
20. Dromey B, Rykovanov SG, Yeung M, Horlein R, Jung D, Gauthier JC, Dzelzainis T, Kiefer D, Palaniyappan S, Shah RC, Schreiber J, Ruhl H, Fernandez JC, Lewis CLS, Zepf M, Hegelich BM (2012) Coherent synchrotron emission from electron nanobunches formed in relativistic laser-plasma interactions. Nature Phys 8(11):804–808
21. Hidding B, Pretzler G, Clever M, Brandl F, Zamponi F, Lubcke A, Kampfer T, Uschmann I, Forster E, Schramm U, Sauerbrey R, Kroupp E, Veisz L, Schmid K, Benavides S, Karsch S (2007) Novel method for characterizing relativistic electron beams in a harsh laser-plasma environment. Rev Sci Instr 78(8):083301
22. Glazyrin IV, Karpeev AV, Kotova OG, Bychenkov VYu, Fedosejevs R, Rozmus W (2012) Ionization-assisted relativistic electron generation with monoenergetic features from laser thin foil interaction. AIP Conf Proc 1465(1):121–127
23. Kiefer D, Henig A, Jung D, Gautier DC, Flippo KA, Gaillard SA, Letzring S, Johnson RP, Shah RC, Shimada T, Fernâš °ndez JC, Liechtenstein VK, Schreiber J, Hegelich BM, Habs D (2009) First observation of quasi-monoenergetic electron bunches driven out of ultra-thin diamond-like carbon (dlc) foils. The Eur Phys J D At Mol Opt Plasma Phys 55:427–432
24. Hu SX, Anthony FS (2002) Gev electrons from ultraintense laser interaction with highly charged ions. Phys Rev Lett 88(24):245003
25. Hu SX, Anthony FS (2006) Laser acceleration of electrons to giga-electron-volt energies using highly charged ions. Phys Rev E 73(6):066502
26. Mulser P, Bauer D, Ruhl H (2008) Collisionless laser-energy conversion by anharmonic resonance. Phys Rev Lett 101:225002
27. Stupakov GV, Zolotorev MS (2001) Ponderomotive laser acceleration and focusing in vacuum for generation of attosecond electron bunches. Phys Rev Lett 86(23):5274–5277
28. Dodin IY, Fisch NJ (2003) Relativistic electron acceleration in focused laser fields after above-threshold ionization. Phys Rev E 68:056402
29. Quesnel Brice, Mora Patrick (1998) Theory and simulation of the interaction of ultraintense laser pulses with electrons in vacuum. Phys Rev E 58(3):3719–3732
30. Popov KI, Bychenkov VYu, Rozmus W, Sydora RD (2008) Electron vacuum acceleration by a tightly focused laser pulse. Phys Plasmas 15(1):013108
31. Popov KI, Bychenkov VYu, Rozmus W, Sydora RD, Bulanov SS (2009) Vacuum electron acceleration by tightly focused laser pulses with nanoscale targets. Phys Plasmas 16(5):053106
32. Wang X, Nishikawa K, Nemoto K (2006) Observation of a quasimonoenergetic electron beam from a femtosecond prepulse-exploded foil. Phys Plasmas 13(8):080702
33. Kluge T, Bussmann M, Gaillard SA, Flippo KA, Gautier DC, Gall B, Lockard T, Lowenstern ME, Mucino JE, Sentoku Y, Zeil K, Kraft SD, Schramm U, Cowan TE, Sauerbrey R (2010) Low-divergent, energetic electron beams from ultra-thin foils. AIP Conf Proc 1209(1):51–54
34. Giulietti D, Galimberti M, Giulietti A, Gizzi LA, Numico R, Tomassini P, Borghesi M, Malka V, Fritzler S, Pittman M, Phouc TK, Pukhov A (2002) Production of ultracollimated bunches of multi-mev electrons by 35 fs laser pulses propagating in exploding-foil plasmas. Phys Plasmas 9(9):3655–3658
35. Hörlein R, Steinke S, Henig A, Rykovanov SG, Schnürer M, Sokollik T, Kiefer D, Jung D, Yan XQ, Tajima T, Schreiber J, Hegelich M, Nickles PV, Zepf M, Tsakiris GD, Sandner W, Habs D (2011) Dynamics of nanometer-scale foil targets irradiated with relativistically intense laser pulses. Laser Part Beams FirstView:1–6
36. Mangles SPD, Walton BR, Tzoufras M, Najmudin Z, Clarke RJ, Dangor AE, Evans RG, Fritzler S, Gopal A, Hernandez-Gomez C, Mori WB, Rozmus W, Tatarakis M, Thomas AGR, Tsung FS, Wei MS, Krushelnick K (2005) Electron acceleration in cavitated channels formed by a petawatt laser in low-density plasma. Phys Rev Lett 94:245001
37. Willingale L, Nagel SR, Thomas AGR, Bellei C, Clarke RJ, Dangor AE, Heathcote R, Kaluza MC, Kamperidis C, Kneip S, Krushelnick K, Lopes N, Mangles SPD, Nazarov W, Nilson PM,

Najmudin Z (2009) Characterization of high-intensity laser propagation in the relativistic transparent regime through measurements of energetic proton beams. Phys Rev Lett 102:125002

38. Pukhov A, Sheng Z-M, Meyer ter Vehn J (1999) Particle acceleration in relativistic laser channels. Phys Plasmas 6(7):2847–2854
39. Gahn C, Tsakiris GD, Pukhov A, Meyer-ter J Vehn M, Pretzler G Thirolf P, Habs D, Witte KJ (1999) Multi-mev electron beam generation by direct laser acceleration in high-density plasma channels. Phys Rev Lett 83(23):4772–4775

Chapter 5
Coherent Thomson Backscattering from Relativistic Electron Mirrors

Having studied the electron dynamics in laser–nanofoil interactions in the previous chapter, we shall now turn our interest to the envisioned application—the intense, short wavelength generation via the reflection of a laser pulse from a relativistic electron mirror.

The experimental realization of the backscatter experiment is demanding for many reasons. First, a high contrast, high intensity laser is required as the driving pulse. Achieving both pulse parameters simultaneously is still a great challenge for state-of-the-art high power laser systems. Second, a powerful probe pulse is needed, which is set up in counter-propagating direction. Achieving good spatio-temporal overlap in the colliding beam configuration, however, is experimentally not trivial. Last but not least, experiments with nanometer thin foils are naturally limited to only a few shots and thus having full control on both pulses in the experiment is crucial and in fact requires accurate preparation of each target shot.

In this chapter, the first experimental study on the generation of a relativistic electron mirror from a nanometer thin foil is presented [1]. Complementary to the experimental results, a complete numerical study on the electron mirror generation and reflection process is given in full depth.

5.1 Experimental Setup

The experiment was conducted at the Astra Gemini dual beam laser facility. The laser system is capable of delivering two optically synchronized laser pulses, which in the following are referred to as the drive and the probe pulse. To cover a broad range of target thicknesses, nanometer foils produced out of two different materials were used in the experiment: (a) carbon foils with thicknesses of 200 nm, 100 nm, 50 nm and density $\rho_C \sim 2.1\,\mathrm{g/cm^3}$, and (b) DLC foils with thicknesses of 25 nm, 10 nm, 5 nm and $\rho_{DLC} \sim 2.8\,\mathrm{g/cm^3}$. To reach the contrast level required for those targets, additional pulse cleaning was applied to the drive pulse. By introducing a re-collimating double plasma mirror into the optical beam path, the contrast of the laser pulse was enhanced to $\sim 10^{-9}$ measured at –2.5 ps prior to the peak of the pulse

© Springer International Publishing Switzerland 2015

D. Kiefer, *Relativistic Electron Mirrors*, Springer Theses,
DOI 10.1007/978-3-319-07752-9_5

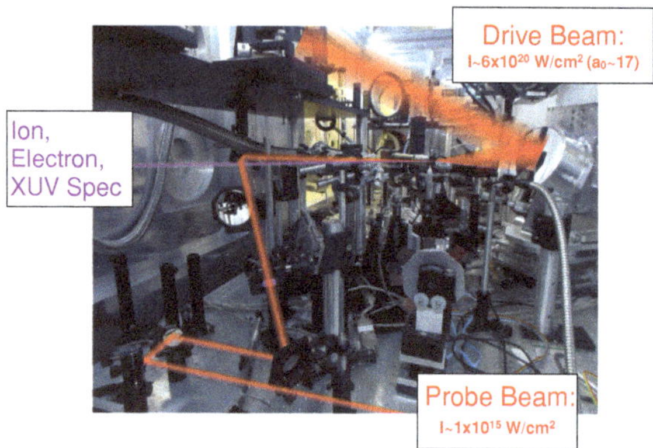

Fig. 5.1 *Photograph of the experimental setup.* The drive pulse is guided on the upper level and focused with a f/2 off-axis parabolic mirror onto the target. The probe beam propagates on the lower level, passes through a focusing lens (f/50) and ascends to a turning mirror, which re-directs the beam to the target. To minimize the angle between drive and probe beam, the turning mirror was carefully positioned in the close vicinity to observation axis of the Multi-Spectrometer, which was set up along the drive beam axis. Only a fraction of the south beam was used as a probe (beam diameter $\sim$17 mm) to ensure a compact probe beam setup relying on 1$''$ optics

(Sect. 3.1.2, Fig. 3.4). Due to the rather low contrast of the probe pulse on the few picosecond time scale ($\sim 10^{-4}$ at -2.5 ps), the peak intensity was set to $\sim 10^{15}$ W/cm^2 such that intensities above the ionization threshold $\sim 10^{12}$ W/cm^2 were reached only a few hundred femtoseconds in advance of the main pulse. To vary the polarization of the drive pulse in the experiment, a $\lambda/4$ wave-plate was positioned in the collimated beam right after the plasma mirror system. The polarization was changed between linear and circular by rotating the wave-plate during the experiment without breaking vacuum.

A photograph of the actual experimental setup is shown in Fig. 5.1 and a schematic illustration of the experimental configuration is given in Fig. 5.2. The drive pulse ($\sim$5 J, 55 fs) is focused with a f/2 off-axis parabolic mirror to a focal spot of 3.5 μm FWHM, reaching peak intensities of 6×10^{20} W/cm^2. Simultaneously, the probe pulse ($\sim$2 mJ, 55 fs) is shot from the opposite side, quasi counter-propagating (angle between both beam axis $\sim 1°$), focused with a f/50 lens to a 55 μm FWHM spot corresponding to a peak intensity of 1×10^{15} W/cm^2.

The radiation emitted from the foil is diagnosed at 0° with respect to the target normal direction using a transmission grating spectrometer. The entrance of the spectrometer was defined by a pinhole with a 200 μm diameter at a distance of 1.3 m, corresponding to a detection angle of 1.7×10^{-8} sr. The transmission grating consists of free-standing gold wires with 1,000 lines/mm, supported by a triangular mesh structure. The backscattered radiation was detected with a micro-channel plate (MCP) that was imaged onto a low noise CCD camera. A detailed description of spectrometer is given in Sect. 3.3.3.

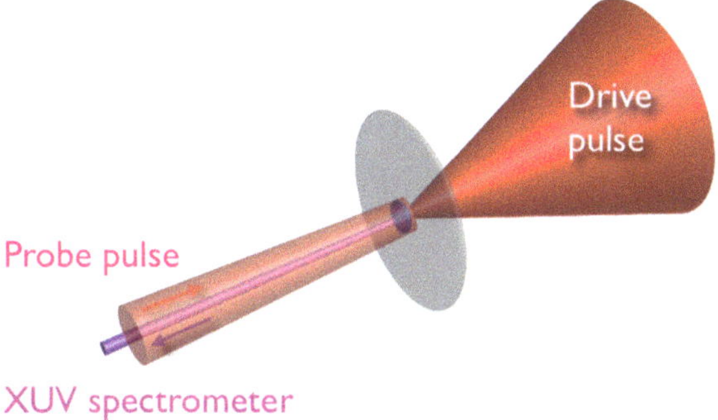

Fig. 5.2 *Schematic illustration of the experimental configuration.* The drive pulse (5 J, 55 fs) is focused to a 3.5 μm FWHM focal spot, corresponding to a peak intensity of 6×10^{20} W/cm^2. The probe pulse (2 mJ, 55 fs) is shot simultaneously from the opposite side (angle between both beam axis ~1°), focused to a 55 μm FWHM spot, which equates to 1×10^{15} W/cm^2. The radiation emitted from the foil is measured at 0° with respect to target normal direction using a transmission grating spectrometer

5.1.1 Spatio-Temporal Overlap

The precise overlap of the drive and probe pulse in space and time is of utmost importance for the backscatter experiment. To relax the requirements on the beam pointing stability and avoid potential jitter problems, the focal spot of the probe pulse was chosen rather large. To achieve spatial overlap, the intersection point of the drive and probe pulse was defined by the tip of a wire (diameter: 7 μm), which both beams were pointed onto, using the high magnification focal spot diagnostic for the drive and a side view imaging system for the probe.

The relative timing of both pulses was determined with the aid of plasma shadowgraphy using an additional transverse probe pulse, schematically shown in Fig. 5.3. Here, the drive pulse was shot at atmospheric pressure at intensity levels well above the limit of air breakdown, which thus caused the formation of a plasma channel in the focal region. Shadowgrams of the generated plasma channel were observed in the transverse probe imaging, once the transverse probe, backlightening the plasma channel, was timed to within the channel's lifetime (~ns). Temporal synchronization of both pulses was achieved tuning the probe pulse to the onset of the plasma channel formation, which could be determined to better than 30 fs. Similarly, the probe pulse was timed prompt relative to the transverse probe by monitoring the plasma channel generated with the focused probe pulse using the transverse probe as the backlighter.

5.2 Experimental Results

As mentioned earlier, the complexity of the experimental setup including the employed, cutting-edge laser system by itself restricts the experimental study to the proof-of-principle. In particular, the probe data presented in the following was measured on a single day, right after aligning both beams on target carefully.

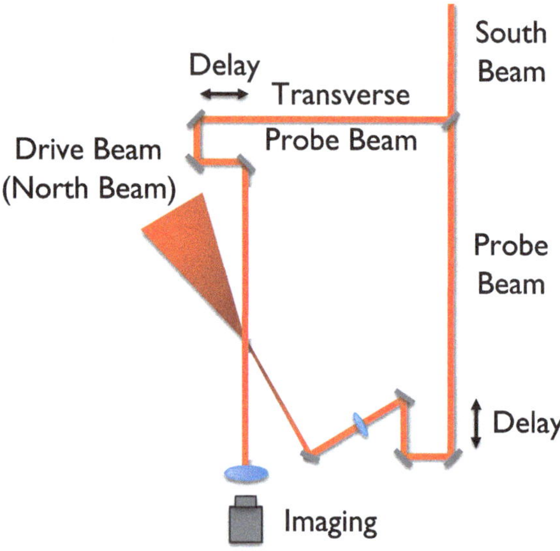

Fig. 5.3 *Pulse synchronization.* Drive and probe pulse were timed relative to the transverse probe beam using plasma shadowgraphy

Figure 5.4a shows the measured photon spectra using the drive laser pulse in linear polarized (LP) configuration. A clear harmonic signal is observed from 200 nm thick carbon foils when irradiating the target with the drive pulse, only. However, reducing the target thickness to 50 and 10 nm thin foils, the harmonic signal breaks down and vanishes in the background noise. This behavior changes substantially when irradiating the target with the probe pulse synchronously. Here, a periodically modulated spectrum ranging down to ~60 nm wavelength was observed repeatedly.

The fundamental difference between the single and dual pulse interaction is evidently seen when directly comparing the raw detector images obtained from two subsequent target shots, shown in Fig. 5.5. Irradiating the target with the drive pulse, only, the signal obtained here is dominated by shot noise. Slight deviations from the background are within the noise level and cannot be attributed to a signal. On the contrary, the subsequent probe shot exhibits a periodically modulated signal, which clearly cannot be explained by any background fluctuations. Although, the signal-to-noise ratio is not ideal due to the small detection angle and could certainly be improved using a collection optic, it is clear from that raw images that the signal is real and obvious to the naked eye. Towards shorter wavelengths, however, the noise level increases and thus prevents detailed analysis.

The measurement was repeated changing the polarization of the incident drive pulse to circular (CP), Fig. 5.4b. In clear contrast to the LP case, the harmonic emission observed from the 200 nm carbon foils is strongly reduced in circular

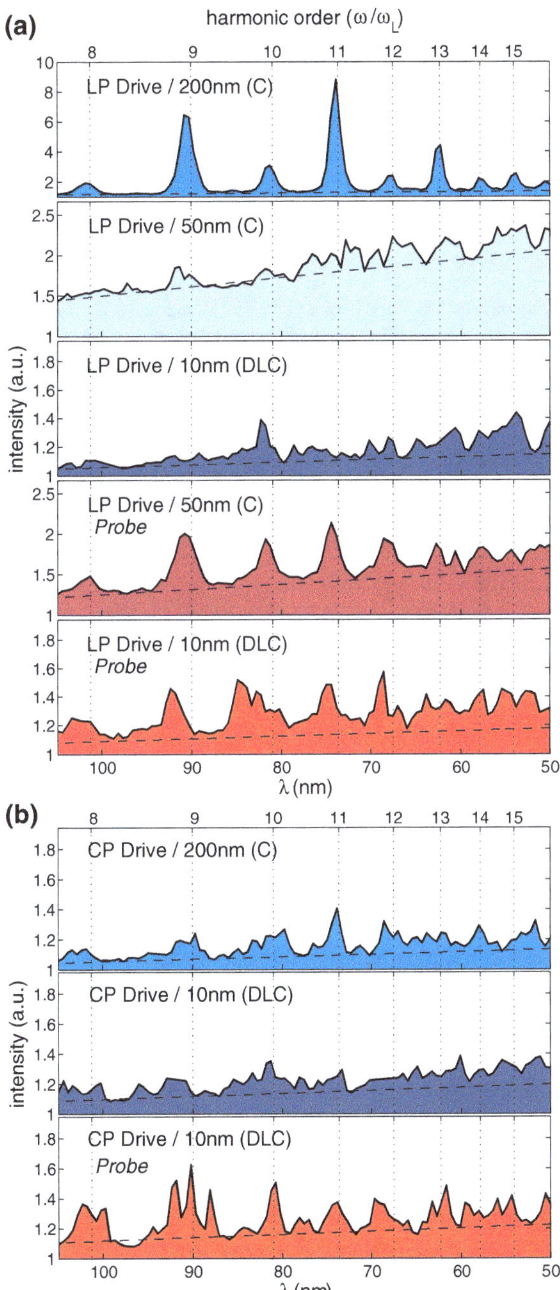

Fig. 5.4 *XUV spectra from different targets and laser pulse configurations.* **a** linear polarized drive (LP), **b** circular polarized drive (CP). The polarization of the probe was linear. Target shots with the synchronized probe pulse are labeled as 'Probe'. *Dashed lines*: linear fit to the background

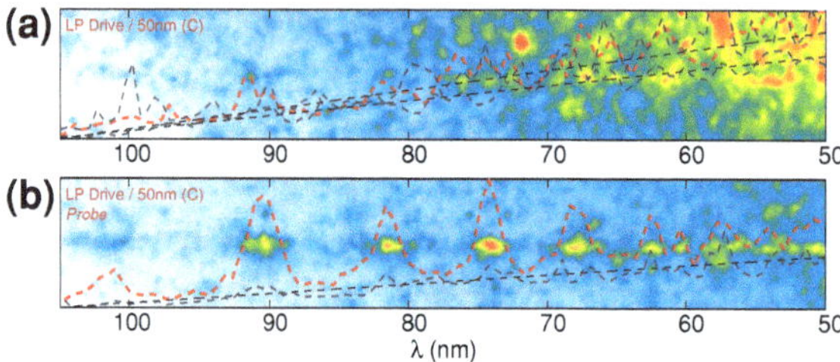

Fig. 5.5 *Detector images* (hot pixels removed) obtained from two subsequent 50 nm target shots, using **a** the drive pulse only, **b** drive and probe pulse simultaneously. *Dashed lines* spectrum (*red*), background spectrum (*gray*), linear fit to the background (*black*)

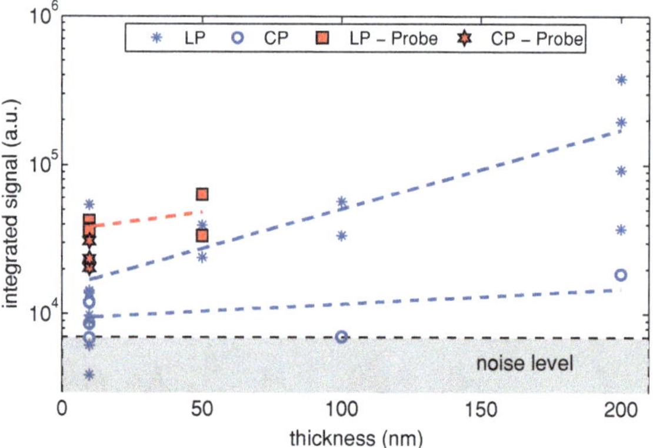

Fig. 5.6 *Signal within* (55–100) nm *from different targets and laser pulse configurations*. Linear fits to the data points are given as a guide to the eye (*dashed lines*)

configuration. Accordingly, no harmonic signal is observed from a 10 nm thin foil, when irradiating the target with the CP drive laser pulse, only. In contrast, a clear backscatter signal was found irradiating the target with both pulses simultaneously.

The experimental observations are summarized in Fig. 5.6, showing the integrated XUV signal measured within 55–100 nm for various interaction configurations. Owing to the complexity of the experiment, the statistics of the experimental data taken is rather limited. Nonetheless, the dataset clearly follows those trends discussed in Fig. 5.4a, b.

Electron Signal

The Multi-Spectrometer allowed measuring simultaneously the emitted XUV radiation and the generated electron distribution in target normal direction. The electron spectra recorded from the same target shots, in which the presented XUV spectra were taken (Fig. 5.4), are shown in Fig. 5.7. The electron distributions observed from single and dual pulse interactions are almost identical. Slight deviations are within

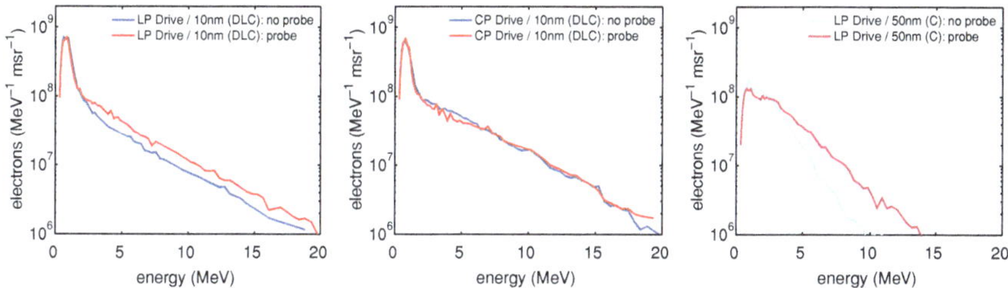

Fig. 5.7 *Electron spectra observed from single and dual beam interactions.* The depicted electron distributions were measured from the same target shots as the photon spectra shown in Fig. 5.4

shot-to-shot fluctuations, whereas the impact of the secondary pulse is negligible. In addition, neither an XUV nor an electron signal was measured when irradiating the foil with the probe pulse, exclusively.

Harmonic Signal

The observation of harmonic radiation in transmission of rather thick (100–200 nm) foils is a new discovery and was for the first time observed at the Astra Gemini laser in this experimental campaign. In fact, this signal was recently attributed to a new generation mechanism, dubbed "Coherent Synchrotron Emission" (Dromey et al. [2]), which is currently under theoretical [3] and experimental [4] investigations. However, this process is inherently different to the coherent backscattering[1] and seems to be efficient only for much thicker targets as compared to the mirror case. The following theoretical analysis will concentrate on the electron mirror generation from laser nanofoil interactions and in particular on the understanding of the observed backscatter signal.

5.3 PIC Simulation

In order to gain deeper insight into the interaction dynamics, two dimensional particle-in-cell simulations were conducted using the PSC code [5]. The simulation of the dual beam configuration in connection with a nanometer thin, solid density plasma is a non-standard PIC simulation and various different tools had to be developed to diagnose the simulation in great detail. Regarding the rather long rise time of a 50 fs gaussian laser pulse, and the high spatio-temporal resolution needed to accurately resolve the mirror structure, and accordingly the back-reflected short wavelength radiation, the simulations carried out in this chapter were computationally expensive.

[1] The relativistic Doppler upshift is non-existent in transmission and thus cannot be attributed to the observed signal.

Table 5.1 *Laser pulse parameter used in the PIC simulation*

	a_0	I (W/cm^2)	τ (fs)		w_0 (μm)	
Drive:	17	6×10^{20}	50	(Gaussian)	3	(Gaussian)
Probe:	0.05	6×10^{15}	100	(Flattop)	50	(Gaussian)

The shape of the field envelope is given in brackets. τ: pulse duration (FWHM), w_0: focal spot size (FWHM)

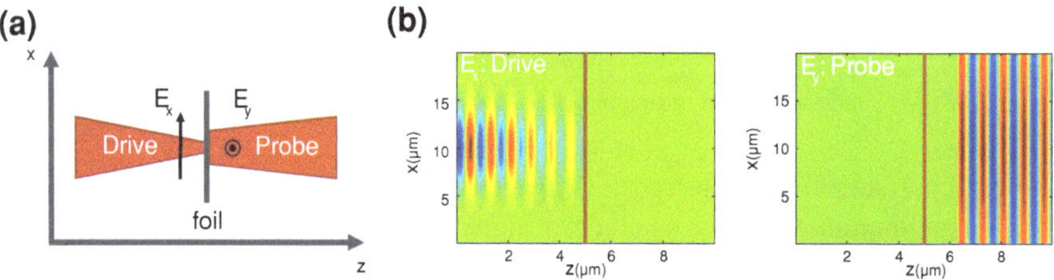

Fig. 5.8 *PIC simulation configuration.* Drive and probe pulse were initialized counter-propagating, in cross-polarized configuration. The plasma layer was positioned at z = 5 µm, the detector recording the electric field at z = 8 µm

In more detail, the simulation box was 10 × 20 µm in longitudinal and transverse dimension, divided into 4,000 × 4,000 cells, which equates to a spatial resolution Δz : 2.5, Δx : 5.0 nm. The nanometer foil was modeled as a fully ionized carbon plasma with density $n_e = 47\, n_c$ and thickness 100 nm (rectangular shape) using 200 particles per cell. Taking into account a target pre-expansion and density reduction due to limited laser pulse contrast conditions, the plasma parameters chosen in the simulation correspond to a solid, 10 nm foil, as used in the experiment.

The laser pulse parameters used in the simulation are summarized in Table 5.1. The drive pulse profile is set to a gaussian in space and time with parameters matched to the experimental conditions. To ensure full overlap of the generated mirror structures and the probe pulse, the temporal profile of the probe beam is set to a flattop shape, probing the whole interaction, whereas the spatial profile is kept as a gaussian. To resolve the field components of the drive and probe pulse independently, the laser pulses were initialized in cross-polarized configuration. In the following, the drive pulse has the electric field component E_x, while the probe pulse is set to E_y, as shown in Fig. 5.8.

In order to monitor the radiation generated during the interaction at the rear side of the foil, a detector was positioned at z = 8 µm, recording the electric field components E_x, E_y within x = 4 µm and x = 16 µm.

In the following, the simulation results are presented in two sections. First, the observed radiation is analyzed in the spectral domain (Sect. 5.3.1). In a second step, the interaction is discussed in the temporal domain in very details (Sect. 5.3.2).

5.3.1 Spectral Analysis

The recorded time-dependent electric fields are Fourier transformed in both polarizations to obtain the spectral intensity as a function of frequency ω and transverse dimension x. Spectral lineouts shown below are obtained averaging in transverse dimension within the spatial region indicated by the dashed lines.

In accordance with the experimental observation, the time-integrated spectrum of the electromagnetic field recorded in transmission of the foil exhibits nearly no signal above $\omega/\omega_L \sim 5$ when irradiating the foil with the drive laser, only (Fig. 5.9a). Redundant, harmonic signal observed off-center stems from target denting late in the interaction, where the laser field is effectively oblique incident on the side wings of the plasma layer. Those harmonic orders show strong dependence on the lateral position x in the spectrum, which is a result of the fact that these harmonics are emitted at a steep angle with respect to the laser axis (thus, pass through the detector at an oblique angle, equivalent to a frequency shift in the spectral domain). Due to the apparent off-axis emission, we do not expect to observe the residual harmonic signal in the experiment, as in stark contrast, the emission was measured on-axis. Moreover, the measured signal is fully confined to the polarization axis of the drive pulse E_x, whereas the signal recorded simultaneously in the polarization axis perpendicular to the drive pulse is governed by computational noise on a much lower signal level, as shown in Fig. 5.9b. In consequence, any signal observed in E_y direction can be unambiguously attributed to the probe pulse.

In contrast to these observations, a clearly modulated spectrum is obtained irradiating the plasma layer with drive and probe pulse synchronously, Fig. 5.10a. The observed signal extends up to $\omega/\omega_L \sim 13$ in excellent agreement with the experimental observation. Moreover, the spectral interference observed in the experimental measurements is clearly visible in the obtained PIC spectrum.

To gain deeper insight, a temporal filter (window function: supergaussian, 40th order) is applied to the recorded electromagnetic field prior to the Fourier transformation, such that the obtained spectrum contains spectral components generated within that time window, only. Shifting that window function in time, the time interval of most efficient back-reflection is identified.

Figure 5.10b shows the spectrum of the time-windowed electric field, t $= [-14, -10]$ T_L. The filtered spectrum now reveals slower spectral decay as the window function truncates time steps where the mirror formation, or reflection is very ineffective. By doing so, we neglect any experimental sophistication such as timing issues. Hence, the filtered spectrum is rather representative to the spectral scaling of the reflection process itself. Moreover, it gives a first hint, that main spectral contributions are generated in the early phase of the interaction, at the time period when the foil is still opaque to the laser, as we shall examine in Sect. 5.3.3 in more detail.

The clean spectrum reveals a systematic dependence of the observed harmonic orders on the lateral position. While odd orders exhibit enhanced signal in the central region, spectral peaks of even order are confined to the off-axis emission. As we shall see in the next chapter, the reflecting mirror structures are created periodically at

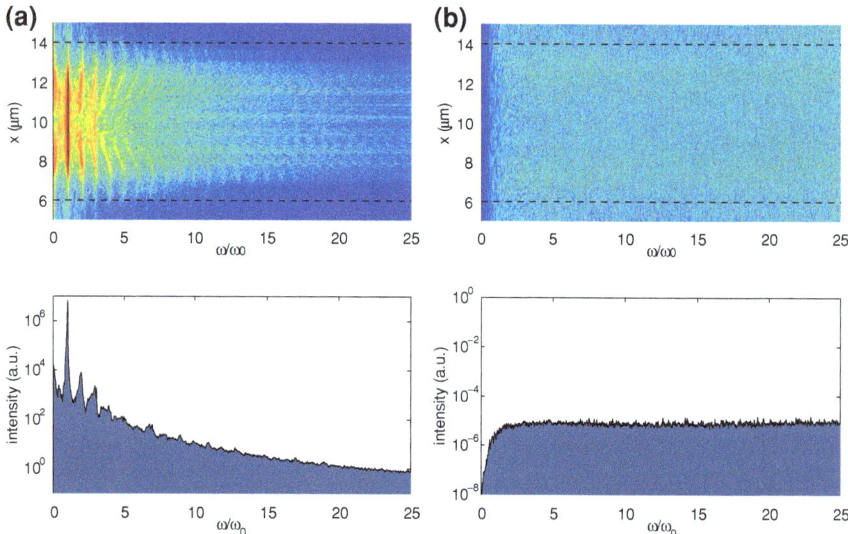

Fig. 5.9 *Spectra obtained from the drive pulse—foil interaction.* **a** Spectrum of the recorded E_x field (transmitted drive laser field), **b** spectrum of the recorded E_y field

twice the laser frequency ($2\omega_L$) due to the driving $v \times B$ force of the laser. However, adjacent electron mirrors are generated with opposite transverse momentum as the electric field, acting on the plasma layer synchronously, oscillates at the laser frequency ω_L. Hence, subsequent electron bunches are ejected in opposite transverse directions, resulting in an overall mirror structure with periodicity of $\lambda_L/2$ in the central, and λ_L in the outer region. Thus, as a direct consequence of Fourier analysis, the periodicity of the harmonic orders observed in the spectrum is $2\omega_L$ in the center (mirror spacing: $\lambda_L/2$), and ω_L in the non-overlapping, outer region (mirror spacing: λ_L). Consequently, the spatially averaged spectra exhibit odd and even harmonics orders, hence a harmonic spacing of ω_L, as observed in the experiment.

5.3.2 Temporal Analysis: Reflection from a Relativistic Electron Mirror

To gain deeper insight into the generation process of the high frequency components observed in the spectrum, the interaction is analyzed in the time domain. Figure 5.11a shows the electron density distribution seen rather early in the interaction, at $t = -15T_L$ relative to the peak of the pulse. At this stage, the periodic generation of attosecond short electron bunches is dominating the electron dynamics. These bunches are created via the driving $v \times B$ force of the laser, acting on the skin layer of the solid density plasma at a frequency of $2\,\omega_L$. As a result, dense electron

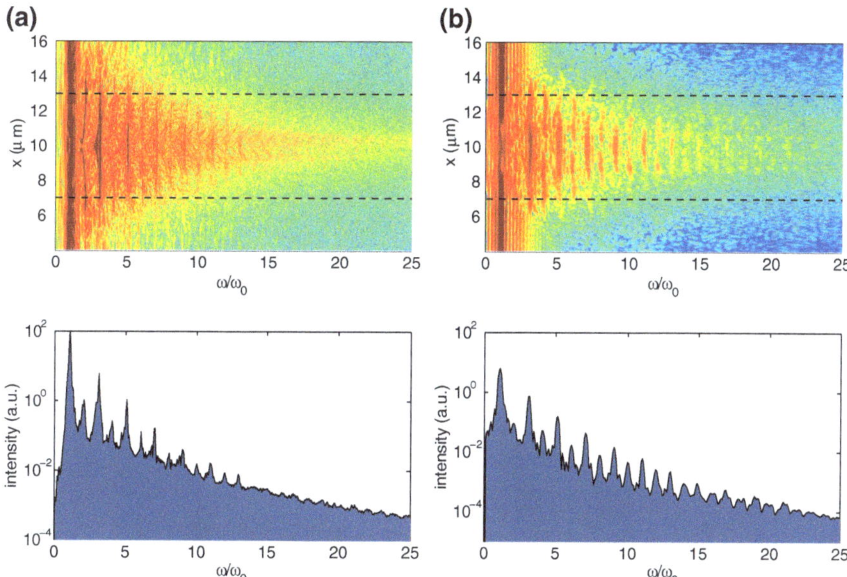

Fig. 5.10 *Spectra obtained from the dual pulse—foil interaction* **a** Spectrum of the backscattered probe field taken from a 50 fs time window. **b** The spectrum filtered with a four cycle time window at the time of most efficient mirror production

bunches are formed at the boundary and accelerated into the plasma, periodically, at every half cycle of the laser field. Each of these bunches traverses the thin plasma quasi-instantaneously, and escapes into vacuum region at the rear side of the plasma as a nanometer thin layer with density well above critical density (i.e. $>10^{21}\,\text{cm}^{-3}$) while propagating in free space at relativistic velocities.

As soon as the electron bunch reaches the rear side of the foil, it encounters the probe field and scatters off radiation. The extremely short length scale of the created relativistic structure ($\sim$10 nm) in connection with its high density ($\sim 3\,n_c$) allows for the coherent scattering, i.e. the mirror-like reflection. In counter-propagating geometry, the scattering amplitudes of the backscattered radiation add up constructively, in direction normal to the mirror surface, and the created electron bunch acts in the coherent case as one expect intuitively from a mirror, that is the radiation is reflected in specular direction. The mirror structure is formed by electrons which are not propagating exactly in the same direction or at the same velocity. However, in the relativistic limit, the velocity dispersion is sufficiently small for electrons of different energies. As a result, the mirror structure remains intact over micron-scale distances, sufficient for the reflection to take place. The relevant γ factor governing the relativistic frequency upshift is $\gamma_z = 1/\sqrt{1-\beta_z^2}$, as discussed in Sect. 2.6. As each mirror constitutes of electrons of various energies (Fig. 5.11b), the backscattered radiation is shifted to a rather broad photon energy range, giving rise to coherent spectral contributions up to a maximal frequency of $\omega_{max} = (1+\beta_{z,max})^2\,\gamma_{z,max}^2\,\omega_L$ from each

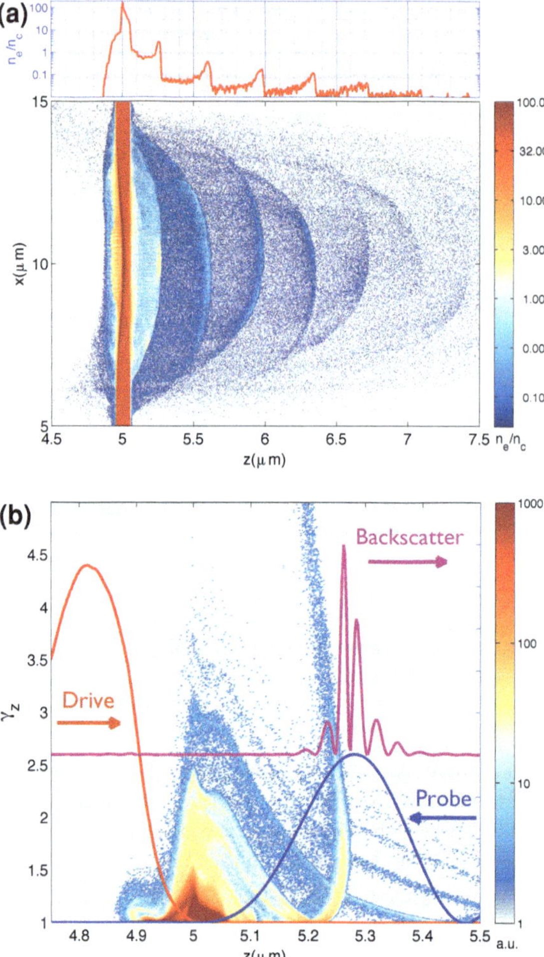

Fig. 5.11 *Reflection from a relativistic electron mirror.* (**a**) A dense electron bunch with thickness ~10 nm FWHM is created at the laser plasma interface by the $v \times B$ force of the driving laser and pushed through the overdense plasma layer. At the rear side of the target, the electron layer escapes from the driving laser field and thus propagates freely in space while reflecting the counter-propagating probe pulse (**b**). The frequency upshift is clearly visible in the backscattered pulse (frequency filter $\omega/\omega_L > 10$)

mirror-like reflection. As the γ_z distribution of the created electron bunches continuously decreases for higher γ_z values, the spectrum of the backscattered radiation slowly merges into the incoherent background rather than dropping off sharply and hence, backscatter signal up to $\gamma_z \sim 2$ is clearly visible above background (Fig. 5.10). This is in good agreement with the experiment, showing a modulated spectrum up to ~65 nm, corresponding to $\gamma_z \sim 1.9$. While each individual electron mirror emits a

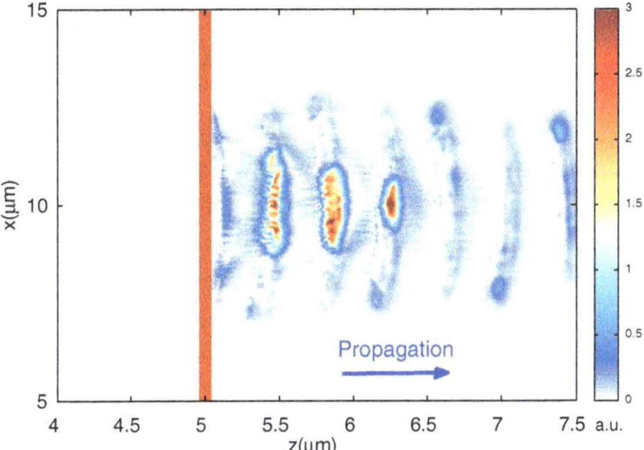

Fig. 5.12 *Backscattered pulse train.* The electron density characteristics of the back-reflecting electron bunches are directly imprinted in the intensity and curvature of the reflected light pulse. For a detailed discussion see Sect. 5.3.4. Note that a frequency filter ($\omega/\omega_L > 5$) was used to visualize the backscatter pulse train

continuous spectrum up to a frequency ω_{max}, the emission from a periodic electron mirror structure results in spectral interference and therefore a strong modulation in the measured photon spectra is observed (as discussed in Sect. 5.3.1).

It is important to note that the counter-propagating probe field passes through the ejected electron mirror, although the layers feature densities above critical density, thus are opaque to an optical wavelength of 800 nm. However, transforming in the rest frame of the mirror the wavelength is $\lambda' = \lambda_L/(1+\beta_z)\gamma_z$ and the mirror density reduces as $n_e' = n_e/\gamma_z$, causing the layer to be partially transparent, as seen in Fig. 5.11b. Thus, the light reflection is the result of the sudden change in density i.e. from vacuum to the electron mirror, analogous to the reflection of optical light from a transparent glass plate.

This reflection process occurs repetitively at every half cycle of the laser field, thus results in a train of attosecond short pulses, as clearly seen in Fig. 5.12. The intensity of each individual pulse is directly correlated to the electron bunch properties they reflect off and is discuss in Sect. 5.3.4 in more detail. Moreover, the emission is directed along the mirror surface normal, as opposed to the emission cone of individual scatterer, which points off-normal, in propagation direction (Sect. 2.6, Fig. 2.10). Thus, the observed high directionality of the emission in specular reflection is a clear signature of the coherence of the scattering process.

5.3.3 Electron Mirror Properties

While electron mirrors are repeatedly generated at every half cycle of the driving laser field, their properties vary significantly over the course of the interaction owing to the complex interplay between the driving laser field and the oscillating plasma layer. Figures 5.13 and 5.14 summarize the electron mirror characteristics observed at different time steps during the interaction.

In the very early phase of the interaction, the created electron layers are extremely well-confined in space, however feature rather low densities and γ_z values, Fig. 5.13a. As the laser pulse rises to higher intensities, both the density and γ_z factor increases, while the sharpness of the electron bunches is still maintained (Fig. 5.13b). The thus generated electron mirrors backscatter the counter-propagating probe pulse most effectively, as clearly seen in Fig. 5.13c. While higher γ_z factors can be observed as the pulse intensity rises closer to the peak, the respective electron bunches become broadened in space, Fig. 5.14a. The degradation of the electron bunch properties can be explained by strong heating of the plasma layer. As the laser intensity rises slowly over many cycles, the plasma electrons go through many oscillations of the driving force. Each of those cycles, however, considerably broadens the electron phase space of the plasma layer. As a result, the electron mirrors created from the bulk plasma loose coherence as time evolves, and so does the backscattered light. Finally, at $t = -8T_L$, the plasma turns transparent to the laser, Fig. 5.14b. In the transparent regime, the generated electron bunches are broad $\sim\lambda_L/2$ (Sect. 4.1), much longer than the wavelength of the reflected light, and therefore the coherent backscatter signal breaks down completely in this phase (Fig. 5.14c).

5.3.4 Electron Mirror Reflectivity

The periodic emission of electron bunches inherently results in a multilayer mirror structure, as observed in Fig. 5.11a. However, the density of each individual electron layer drops comparably fast as it propagates in vacuum ($\sim$ one order of magnitude within a distance of $\sim\lambda_L/2$). Taking into account that the reflectivity is expected to scale with $R_m \propto n_e^2$ (Sect. 2.5.1, Eq. 2.42), we can neglect potential contributions from multiple reflections and discuss the reflection process from isolated bunches in the region of their highest density, that is in the vicinity of the target rear side. As a result, we can relate each back-reflected pulse to an electron bunch, which it originates from. As the electron bunch parameters vary significantly over the course of the interaction, we can gain deeper insight into the reflection process and identify how different bunch parameter affect the electron mirror reflectivity.

To deduce the reflectivity of a single electron bunch at a certain wavelength of the back-scattered radiation, we apply a spectral filter to the electron distribution of the acting mirror and the electric field of the backscattered radiation. The electric field is frequency filtered within $9\omega_L < \omega < 11\omega_L$ and the peak intensity of the backscattered pulse is extracted from the resulting intensity distribution.[2] The electron density of the corresponding electron bunch is filtered in phase space such that

[2] At $\omega = 10\,\omega_L$, a minimal bandwidth of $\Delta\omega/\omega = 20\,\%$ is needed to resolve different pulses spatially.

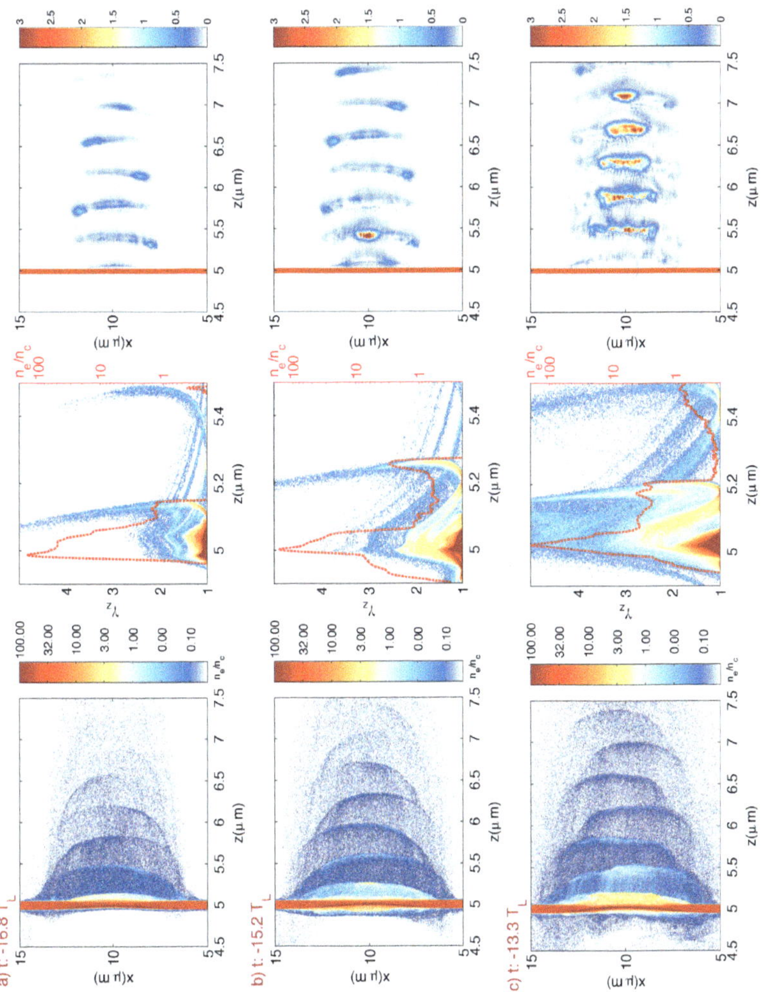

Fig. 5.13 *Electron mirror properties: Time series I.* **(a)** at early time steps, the created electron bunches have rather low density and γ_z values and thus the backscatter signal is low. As the pulse rises in intensity, the density and the γ_z factor of mirror structures increase and thus a strong, clean backscatter signal is observed **(b)**, **(c)**. The probe field is frequency filtered ($\omega/\omega_L > 5$) and is shown slightly later in time ($\sim T_L/2$)

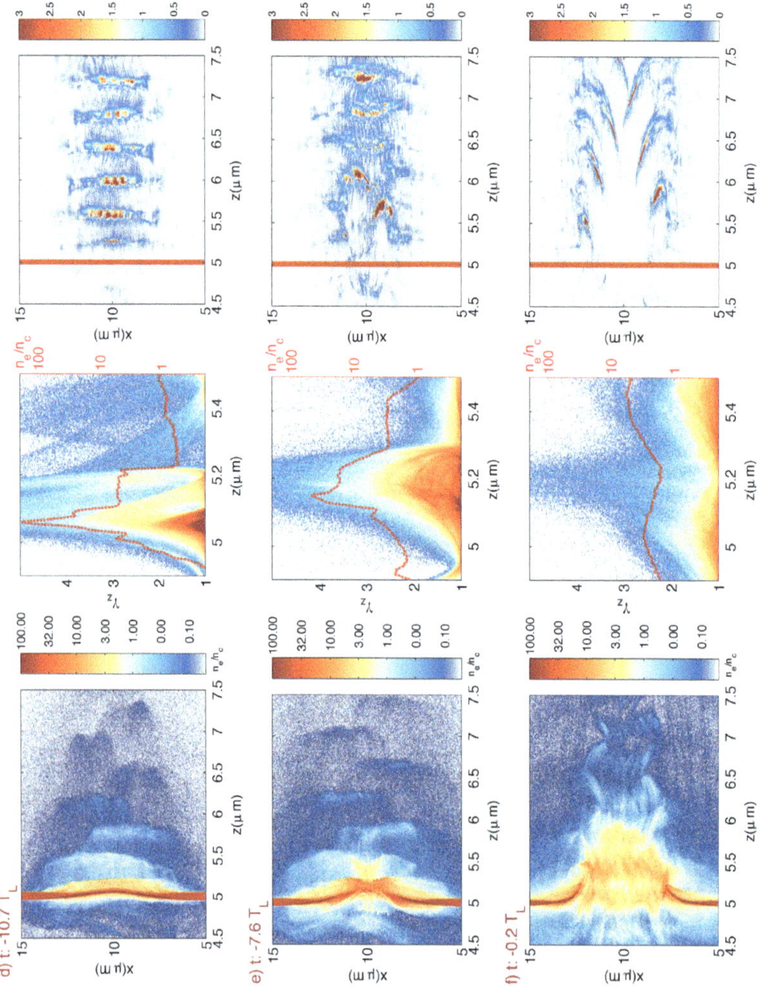

Fig. 5.14 *Electron mirror properties: Time series II.* **(d)** closer to the peak of the driving pulse, the plasma layer deforms, strong electron heating is observed and the backscattered pulses become increasingly more distorted. As the plasma layer turns transparent **(e)**, the electron bunches broaden significantly and therefore the backscattered signal breaks down completely in this phase **(f)**

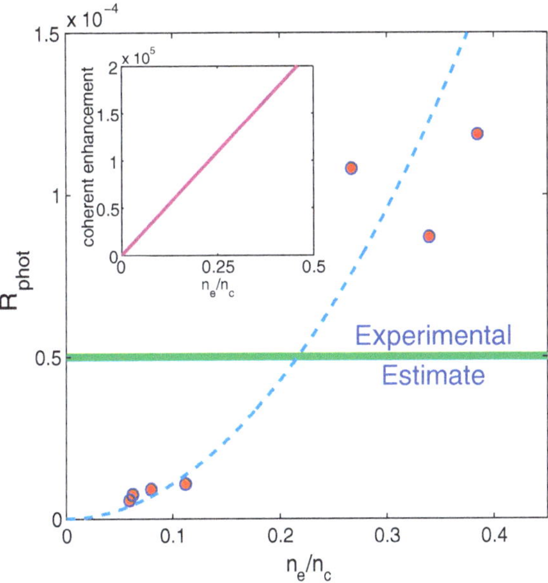

Fig. 5.15 *Electron mirror reflectivity*. The reflectivity obtained from PIC simulation (*dots*) follows a quadratic scaling as expected from coherent scattering theory (*analytic curve*). The electron mirror reflectivity deduced from the experiment (*green*) is 5×10^{-5}, in fair agreement with the expected value. Inset: coherent enhancement R_m/R_{incoh} as expected from theory (Sect. 2.5.1, Eqs. 2.42 and 5.1)

the remaining electrons all satisfy $9 < (1+\beta_z)^2\gamma_z^2 < 11$ and the peak density of the resulting monochromatic electron mirror is extracted. The ratio of incident and reflected intensity deduced from the simulation relates to the mirror reflectivity as $I_r/I_i = (1+\beta_z)^4\gamma^4 R_m$ (Sect. 2.5.2, Eq. 2.44), from which we calculate the mirror reflectivity R_m.

For the sake of simplicity, we focus the analysis on the early phase of the interaction, where the electron bunch properties and the backscattered pulses reveal smooth behavior, and only the bunch density varies significantly for different bunches (Fig. 5.13).

The result is shown in Fig. 5.15. We clearly recover the quadratic behavior, characteristic for the coherent emission. Moreover, the analytical curve, derived in Sect. 2.5.1, fits well to the extracted PIC data, using electron bunch parameters deduced from the simulation. The importance of the coherence of the scattering process observed in the simulation becomes even more evident, comparing the signal to the incoherent scattering. Using identical bunch properties ($n_0 = 0.1 - 0.3\, n_c$, $d = 10\,\mathrm{nm}$), we expect for the incoherent electron bunch reflectivity

$$R_{incoh} = \frac{\sigma_T}{A} N = \sigma_T \sqrt{\pi} n_0 d = (2-7) \times 10^{-10} \tag{5.1}$$

Clearly, such an inefficient process is unlikely to be measured in the experiment. To be more accurate, we shall go back to the experiment and make a rough estimate on the photon number of the backscattered pulse measured in the experiment.

5.3.5 Photon Number Estimate

In the following, the absolute photon number measured at $\lambda = 80\,\text{nm}$ is deduced from the signal count level on the detector, taking into account the ratio of the solid angle resolved in the measurement Ω_{spec}, relative to the expected angular distribution of the emitted radiation Ω_{total}:

$$N_{total}^{photon} = N_{spec}^{photon} \times \frac{\Omega_{total}}{\Omega_{spec}}. \tag{5.2}$$

The cone angle of the backscattered radiation can be estimated, assuming a diffraction limited light cone with apex angle 2α determined by the source size with diameter $2w_0$

$$2\alpha = \theta \sim \frac{2}{\pi}\frac{\lambda}{w_0} \quad \text{and} \quad \Omega_{total} = 4\pi\,\sin^2(\alpha/2). \tag{5.3}$$

where w_0 is the radius at which the intensity drops down to I_0/e^2 and is related to the FWHM diameter d_{FWHM} as $w_0 = d_{FWHM}/\sqrt{2\ln 2}$. Using no collection optic, the solid angle resolved by the transmission grating spectrometer is determined by the diameter of the entrance pinhole $D_2 = 200\,\mu\text{m}$ and its distance to the source: 1350 mm, thus $\Omega_{spec} = 1.7 \times 10^{-8}$ sr.

In the experiment, the signal observed on the detector within the spectral peak at $\lambda \sim 80\,\text{nm}$ ($\Delta\lambda/\lambda \sim 10\,\%$) was on the few photon count level (Fig. 5.4). Taking into account the efficiency of the MCP: $\eta_{MCP} \sim 0.1$ and the grating: $\eta_{TG} \sim 0.1$, we estimate for the photon number measured at that wavelength with the detector $N_{spec}^{80nm} \sim 200$ photons as a rather conservative value. Note, that a bandwidth of 20 % was assumed here to be able to compare directly to the PIC simulation results. From the simulation, we deduce that the size of the electron mirrors is determined by the central region of the drive laser focus, $\sim 2\,\mu m$ (Fig. 5.13), corresponding to an emission into a solid angle of $\Omega_{total} = 7 \times 10^{-4}$ sr. Hence, we estimate $N_{total}^{80nm} \sim 8 \times 10^6$ photons/shot at a wavelength of 80 nm within a bandwidth of 20 %.

Moreover, the lifetime of the mirror is of the order of half an optical cycle. Accordingly, we estimate that 1.6×10^{11} probe photons with $0.8\,\mu\text{m}$ wavelength interact with the mirror, which equates to a mirror reflectivity of $\sim 5 \times 10^{-5}$.

Indeed, the estimated reflectivity is in good agreement with the reflectivity deduced from the PIC simulation (Fig. 5.15) and can only be understood taking into account the coherent enhancement of the Thomson cross-section causing a drastic increase in the backscattered signal by at least four to five orders of magnitude.

References

1. Kiefer D, Yeung M, Dzelzainis T, Foster PS, Rykovanov SG, Lewis CLS, Marjoribanks RS, Ruhl H, Habs D, Schreiber J, Zepf M, Dromey B (2013) Relativistic electron mirrors from nanoscale foils for coherent frequency upshift to the extreme ultraviolet. Nat Commun 4:1763
2. Dromey B, Rykovanov SG, Yeung M, Horlein R, Jung D, Gauthier JC, Dzelzainis T, Kiefer D, Palaniyappan S, Shah RC, Schreiber J, Ruhl H, Fernandez JC, Lewis CLS, Zepf M, Hegelich BM (2012) Coherent synchrotron emission from electron nanobunches formed in relativistic laser-plasma interactions. Nat Physi 8(11):804–808
3. An der Brugge D, Pukhov A (2010) Enhanced relativistic harmonics by electron nanobunching. Phys Plasmas 17(3):033110
4. Yeung M, Dromey B, Cousens S, Dzelzainis T, Kiefer D, Schreiber J, Bin H, Ma JW, Kreuzer C, Meyer-ter Vehn J, Streeter MJV, Foster PS, Rykovanov S, Zepf M (2014) Dependence of laser-driven coherent synchrotron emission efficiency on pulse ellipticity and implications for polarization gating. Phys Rev Lett 112:123902
5. Bertsch GF, Bonitz M, Filinov A, Filinov VS, Lozovik Y, Semkat D, Ruhl H (2006) Introduction to computational methods in many body physics. Rinton Press, Princeton

Chapter 6
Conclusions and Outlook

6.1 Summary of the Results

In this work, the relativistic electron dynamics in high intensity laser–nanofoil interactions are investigated in a series of experiments in combination with numerical studies. The presented results advance the understanding of light matter interactions and will help creating novel electron, ion and X-ray beams from the interaction of a high intensity laser with a nanoscale plasma.

In a first set of experiments, the electron distributions created from laser–nanofoil interactions have been studied in great detail with different high intensity lasers, covering nearly the complete range of the currently available high power laser systems. As the target thickness was varied from the thick, micrometer to the nanometer scale two different regimes were found to exist.

In the thick target range, energetically broad, exponentially decaying electron distributions were observed showing rather low dependence on target thickness and good agreement with the theoretical scaling laws, predicting the electron mean energy of the generated hot electron distributions as a function of laser intensity. By reducing the target thickness to the nanometer scale, however, significant increase in the spectrally resolved electron mean energies was found, while on the contrary, the observed ion energies dropped considerably. Both observations were explained by the onset of plasma transparency supported by transmission measurements and numerical simulations. This experimental work constitutes the first comprehensive study on the hot electron generation in high intensity laser–nanofoil interactions and thereby sheds light on fundamental problems in laser solid plasma interactions such as the longstanding question of laser energy absorption.

The reduction in target thickness to the very extreme of ≤ 5 nm thin foils led to the discovery of a new acceleration mechanism (Kiefer et al. [1]), not predicted by any theoretical work prior to the experimental investigations. Here, quasi-monoenergetic electron beams were observed for the first time from ultrathin foils at the MBI and LANL laser system peaked in the energy distribution at $\sim$4 and $\sim$35 MeV, respectively. The observed electron energies are remarkably close to those expected from

© Springer International Publishing Switzerland 2015

D. Kiefer, *Relativistic Electron Mirrors*, Springer Theses,
DOI 10.1007/978-3-319-07752-9_6

the free electron scaling, suggesting that the massive reduction in target thickness allows the electrons to be effectively injected from the semi-transparent plasma layer into the transmitted laser field. This observation has attracted great interest in the field and has recently triggered further theoretical investigations.

With regard to the generation of a relativistic mirror structure, the electron dynamics in laser–nanofoil interactions using the existing multi-cycle high power laser pulses turns out to be fundamentally different from the envisioned relativistic electron mirror generation with highly idealized, step-like laser pulses. Nonetheless, in the framework of this thesis, it was demonstrated for the first time, that dense electron mirrors can in fact be created from nanoscale foils (Kiefer et al. [2]).

To investigate the backscattering of a secondary pulse from the relativistic electron bunches generated in laser–nanofoil interactions a dual beam experiment was conducted at the Astra Gemini laser system. Irradiating 10 and 50 nm thin foils with a high intensity drive pulse and a rather weak, counter-propagating probe pulse synchronously, a periodically modulated spectrum ranging down to ~60 nm was observed.

Numerical studies well adapted to the experimental configuration show good agreement with the experimentally observed photon spectra. The simulation suggests that relativistic electron bunches of high density (~5 n_c) and extremely short length-scale (~10 nm) are generated by the driving laser field while the plasma layer is still opaque to the laser. Those extreme properties of the created, freely propagating relativistic structures indeed allow for a mirror-like reflection shifting the frequency of the counter-propagating laser coherently from the visible to the XUV.

The reflection process in combination with the frequency upshift was analyzed in the PIC simulation in very detail. It was shown that the frequency upshift is governed by an effective gamma factor of the collective mirror structure, which is determined by the velocity component normal to the mirror surface $\gamma_z = (1 - \beta_z^2)^{-1/2}$, as opposed to the gamma factor of each individual electron. Moreover, the spectral modulations observed in the backscattered signal were explained by the periodic emission from multiple electron mirrors repetitively created during the rather slow rise of the laser pulse. The mirror reflectivity is seen in simulation to scale quadratically with the number of electrons involved in the reflection process and was well explained analytically in the framework of coherent scattering theory.

The signal observed in the experiment is estimated to be 8×10^6 photons/shot at 80 nm wavelength corresponding to a mirror reflectivity of 5×10^{-5}. This is in good agreement with the signal level expected from PIC simulation and exceeds the signal expected from incoherent Thomson scattering by more than four orders of magnitude. Taken together, while not directly measured, the signal level of the backscattered pulse, the frequency upshift governed by $\sim 4\gamma_z^2$, as well as the periodic modulation in the backscattered signal are strong indications for the observation of a coherent process, i.e. the reflection form a relativistic electron mirror.

The presented work investigates for the first time the generation of a relativistic electron mirror from a nanoscale foil in experiment and in simulation using realistic laser pulse parameters. It is a major achievement of this work to close the vast spanning gap between theoretical ideas and experimentally accessible concepts. The obtained results will give a clear guidance for future developments of a relativistic mirror that could, on a micro-scale, produce bright bursts of X-rays.

6.2 Future Perspectives

6.2.1 Relativistic Electron Bunches from Laser–Nanofoil Interactions

While the creation of a single, solid density, relativistic electron bunch from a few nanometer thin foil crucially relies on few cycle high power laser pulses and still is somewhat beyond of what can be achieved today, the electron bunch generation from solid density foils using current laser technology already is extremely useful and deserves further experimental investigations.

The characteristics of the electron bunches that can be created from solid plasmas are outstanding. PIC simulations as well as first experimental studies (such as the one presented in this thesis) show that electron bunch lengths on the few nano metre scale (hence attosecond short) and densities $>10^{21}$ cm^{-3} can be achieved with current laser technology. These electron bunch properties are unique in many ways and by no means accessible from the "conventional" laser wakefield acceleration mechanism. Although the electron acceleration from underdense plasmas has already proven to be useful to generate incoherent XUV or even X-ray radiation, the electron bunch properties obtained from those interactions are not sufficient to reach the coherent limit.

On the contrary, nanoscale bunches observed from laser solid plasma interactions are ideal for the generation of coherent short wavelength generation. A good example demonstrating the great potential of those bunches is the harmonic emission from nanoscale targets that has recently gain high interest. The first experiments investigating the emission from laser–nanofoil interactions were performed at the MBI, LANL and Astra laser facilities complementary to the work presented in this thesis. Strong harmonic signal was observed in the experiments from rather thick 100–200 nm targets in transmission and normal incidence interaction configuration, which could not be explained by any of the well-established generation mechanisms, such as the relativistically oscillating mirror (ROM) or coherent wake emission (CWE). It was rather found from simulation that the observed emission is due to the formation of extremely dense electron bunches, which, as they perform rapid elliptical orbits at the front side plasma interface, emit synchrotron radiation, coherently [3, 4]. While the periodic electron bunch formation at the plasma vacuum boundary is intrinsic to the interaction with a solid density plasma, it was shown in a parametric study that the plasma length scale crucially affects the electron bunch properties and thus determines the regime of coherent emission [5].

Yet, at present, the theoretical understanding of the electron bunch formation process at the plasma vacuum boundary is very little and it will be a key challenge for future theoretical investigations to give a deeper insight into that process and predict optimal conditions for the effective bunch generation. The measurements performed within this thesis will provide a benchmark for theoretical investigations.

Apart from creating dense relativistic structures to generate coherent burst of XUV or X-ray radiation, the acceleration of electrons from nanoscale targets has the prospect of generating unprecedented high flux electron beams, significantly higher than what is observed from gaseous targets. To date, monoenergetic electron beams are routinely produced from laser wakefield acceleration in underdense plasmas. Those beams show high quality, can be controlled to a high degree and have proven very useful in applications. The electron energies are steadily increasing to beyond the GeV level, however, very little is done to achieve higher particle flux. In fact, the number of electrons that can be accelerated by the laser wakefield mechanism is somewhat limited, owing to the fact that the driving plasma wave becomes easily perturbed by the accelerated particle bunch trapped therein (beam loading, see reference in [6]). On the contrary, it is clear that electron beams from solids could potentially yield high electron currents. So far, little attention was paid to that route as the electron spectra observed from solid plasmas exhibit rather low energetic, exponentially decaying electron distributions. Quasi-monoenergetic distributions at the tens of MeV energy level as observed here from nanoscale foils, however, could pave the way for a novel laser-driven, high current electron source.

6.2.2 Relativistic Electron Mirrors: Towards Coherent, Bright X-rays

The experiment and simulation presented in Chap. 5 of this thesis provides unprecedented deep insight into the scheme of short wavelength generation via coherent Thomson backscattering from relativistic electron mirrors. The emphasis of the investigations presented here is on the proof-of-principle rather than on demonstration of a source ready to use in applications. Nonetheless, we shall discuss the steps necessary to take in the future to achieve shorter wavelengths as well as to increase the signal level and thus move the generation scheme from the proof-of-principle to a versatile source of coherent X-rays.

In backscatter experiments, a rather straightforward way to increase the signal level of the generated short wavelength radiation is to increase the intensity level of the incident probe pulse (or, in case of perfect spatio-temporal overlap, the probe pulse energy). This can be done trivially up to the threshold where the probe field becomes a significant perturbation to the electron bunch dynamics. This threshold is expected at $a_0 \sim 1$, which marks the transition to nonlinear Thomson scattering [7] and thus in the experimental configuration presented in Chap. 5, would allow for an increase in photon number by $\sim 10^3$.

Increasing the electron mirror reflectivity (hence the efficiency of the process), requires the creation of relativistic electron mirrors of even higher density. Owing to the complexity of the bunch formation process at the laser plasma interface, the scaling of the bunch properties with laser and plasma parameters is not clear.

Ultimately, the utilization of extremely sharp rising few cycle laser pulses seems of utmost importance to avoid a strong perturbation or even expansion of the plasma layer prior to the mirror formation. Using such pulses the transition to the envisioned ideal REM generation scenario observed in simulation from step-like rising pulses is expected. Those pulses could give rise to REMs of almost solid densities while maintaining the initial thickness of only a few nano metre. In that case, the bunch density could be increased by a factor of 100, which in turn would boost the reflectivity by 10^4. Such a performance would certainly be outstanding.

To access shorter wavelengths the gamma factor of the electron mirror structure must be increased, which certainly can be realized to some extend by increasing the intensity of the driving laser pulse. A more efficient way would be to achieve that the gamma factor of the mirror structure is identical to the gamma factor of each individual electron forming the mirror structure. This is generally not the case for laser-driven electron mirrors as the transverse field character of the driving pulse imposes transverse momentum to the accelerated electrons, which considerably reduces the effective gamma factor of the mirror structure $\gamma_z = \gamma/\sqrt{1 + p_\perp^2}$ (Sect. 2.6).

Recently, Wu et al. [8] showed that in the transparent regime (Sect. 2.3.2) this major drawback of laser generated electron mirrors can be overcome using a secondary reflector foil. In that scheme, the electron mirror is born from the first nm thin foil, using an intense few cycle laser pulse and accelerated to high energies while surfing on the electromagnetic wave. Upon the reflection from the secondary foil, the driving field separates from the high energetic electron bunch, which passes through the foil. From the conservation of the canonical momentum ($p_\perp - a = const$) one can immediately see that as the electron bunch traverses the reflector foil and separates from the driving field ($a_0 = 0$) the transverse momentum vanishes to zero. As a result, relativistic electron mirrors freely propagating with constant gamma factor and zero transverse momentum are obtained. These electron mirrors were shown to provide a narrowband frequency shift $4\gamma^2\omega_L$ and thereby act close to the originally described relativistic mirror. It was demonstrated in PIC simulation that from such a double foil backscatter scenario, intense X-ray pulses (1 keV, < 10 as, > 10 GW) could be generated, in principle [7]. Yet, the experimental realization clearly relies on the next generation of high power few cycle laser systems ($a_0 \sim 40$) and thus will be subject to experimental investigations in the years to come.

References

1. Kiefer D, Henig A, Jung D, Gautier DC, Flippo KA, Gaillard SA, Letzring S, Johnson RP, Shah RC, Shimada T, Fernâ š°ndez JC, Liechtenstein VK, Schreiber J, Hegelich BM, Habs D (2009) First observation of quasi-monoenergetic electron bunches driven out of ultra-thin diamond-like carbon (dlc) foils. Eur Phys J D 55:427–432
2. Kiefer D, Yeung M, Dzelzainis T, Foster PS, Rykovanov SG, Lewis CLS, Marjoribanks RS, Ruhl H, Habs D, Schreiber J, Zepf M, Dromey B (2013) Relativistic electron mirrors from nanoscale foils for coherent frequency upshift to the extreme ultraviolet. Nat Commun 4:1763

3. Dromey B, Rykovanov SG, Yeung M, Horlein R, Jung D, Gauthier JC, Dzelzainis T, Kiefer D, Palaniyappan S, Shah RC, Schreiber J, Ruhl H, Fernandez JC, Lewis CLS, Zepf M, Hegelich BM (2012) Coherent synchrotron emission from electron nanobunches formed in relativistic laser-plasma interactions. Nat Phys 8(11):804–808
4. Yeung M, Dromey B, Cousens S, Dzelzainis T, Kiefer D, Schreiber J, Bin H, Ma JW, Kreuzer C, Meyer-ter Vehn J, Streeter MJV, Foster PS, Rykovanov S, Zepf M (20147) Dependence of laser-driven coherent synchrotron emission efficiency on pulse ellipticity and implications for polarization gating. Phys Rev Lett 112:123902
5. An der Brugge D, Pukhov A (2010) Enhanced relativistic harmonics by electron nanobunching. Phys Plasmas 17(3):033110
6. Esarey E, Schroeder CB, Leemans WP (2009) Physics of laser-driven plasma-based electron accelerators. Rev Mod Phys 81:1229–1285
7. Wu HC, Meyer-ter Vehn J, Hegelich BM, Fernández JC (2011) Nonlinear coherent thomson scattering from relativistic electron sheets as a means to produce isolated ultrabright attosecond x-ray pulses. Phys Rev ST Accel Beams 14:070702
8. Wu HC, Meyer-ter Vehn J, Fernández J, Hegelich BM (2010) Uniform laser-driven relativistic electron layer for coherent thomson scattering. Phys Rev Lett 104(23):234801

Appendix A
Plasma Mirrors

A plasma mirror (PM) is an ultrafast optical shutter, rapidly changing its optical properties from almost perfectly transmissive to highly reflective. Here, an intense laser pulse is focused onto an anti-reflective coated substrate, which ionizes and forms an overcritical plasma surface at the leading edge of the main pulse and thereby separates the reflected high intensity peak from the pulse preceding low intensity background (Fig. A.1).

To ensure high reflectivity as well as proper triggering of the PM substrate, the fluence on the PM has to be adapted to the initial contrast of the laser system. If the fluence is chosen too high, the plasma forms very early and thus reflects off unwanted signal. Hence, the cleaning effect is rather low. Moreover, a rather long expansion of the plasma surface prior to the reflection of the peak pulse eventually induces wave front distortions and therefore reduced focusability of the reflected beam. In contrast, if the fluence is chosen too low, the PM ionizes too late (or not at all), which in turn reduces the overall reflectivity and energy throughput of the system. Numerous experimental studies show that for a conventional CPA laser system with moderate intrinsic contrast, PMs should be operated in the range of 10–100 s J/cm^2 [1–4]. Moreover, in the case of an oblique incidence configuration, higher reflectivity values are observed from s-polarization owing to an increased energy loss from the resonant absorption mechanism, which becomes important in p-configuration [5]. All in all, for optimized conditions, PM reflectivities up to $\sim$80 % were observed. The contrast enhancement that is achieved is simply determined by the ratio of the plasma reflectivity and the reflectivity of the anti-reflective coating $R_{Plasma}/R_{AR} \sim 10^2$ and can be increased by cascading several PMs and using multiple reflections [6, 7].

In experiment, there are essentially two different ways a PM can be set up. A rather simple implementation is to set up the PM in the target chamber, in the focusing beam of the final off-axis parabolic mirror, directly in front of the target. This scheme can be realized very quickly as it does not require any additional optics or heavy engineering. As part of this PhD work, such a PM system was designed and implemented at the Trident laser system using two PM reflections (Fig. A.2). This system allowed for the first laser shots on nanoscale foils at the Trident laser facility and already

© Springer International Publishing Switzerland 2015

D. Kiefer, *Relativistic Electron Mirrors*, Springer Theses,
DOI 10.1007/978-3-319-07752-9

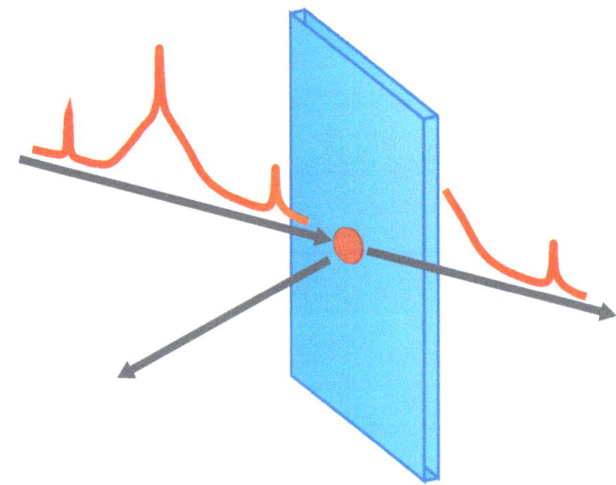

Fig. A.1 *Plasma mirror working principle.* The preceding, low intensity part of the pulse is transmitted through the plasma mirror substrate, whereas the main pulse ionizes the surface and reflects off the plasma

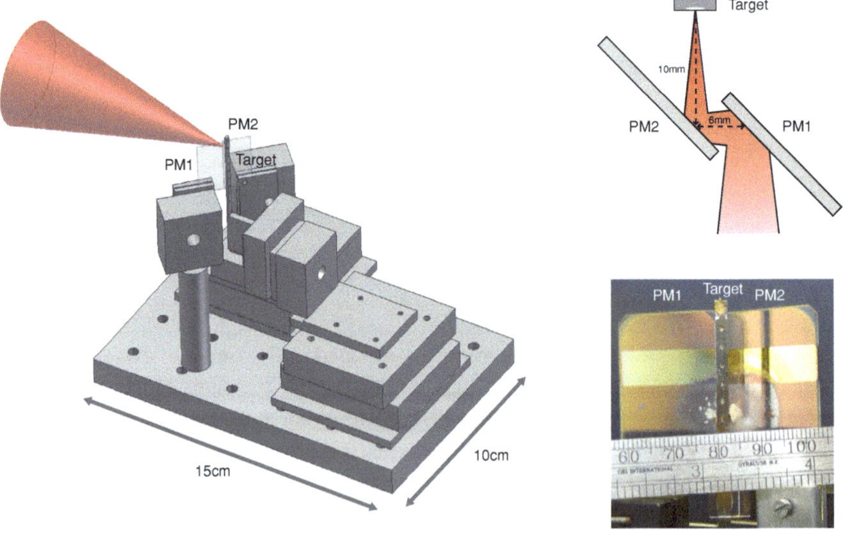

Fig. A.2 *Double plasma mirror setup used at the Trident laser.* The intensity on the plasma mirrors was 5×10^{14} W/cm^2 (PM1) and 2×10^{15} W/cm^2 (PM2), respectively. Plane glass substrates (BK7) coated with an anti-reflective coating ($R < 0.5$ %) were used as PMs. Slim gold stripes on the PMs were employed to facilitate the alignment of the DPM system

demonstrated unprecedented high C^{6+} ion cutoff energies at that time [8]. However, while very successful at the Trident laser, this PM setup is impractical for the use at rather low energy (~1 J) laser systems such as the ATLAS laser. Here, the fluence required to trigger the PM is reached only in the very close vicinity of the target

($\sim$1 mm) due to the factor of $\sim$100 less energy in the beam, making the alignment of the PMs and the precise target positioning impossible. A way to overcome this problem is to decouple the PM from the target interaction and to use rather slow focusing optics, which reach sufficient fluence values already a few centimeters in front of the focal point. Such a re-collimating PM system was built for the ATLAS laser and is described in very detail in the following chapter.

A.1 ATLAS Plasma Mirror

At the MPQ, the great importance of laser pulse contrast with regard to the acceleration of ions was already studied in 2004 [9] and a few years later, the transition from micrometer scale targets to nanometer thin foils made the substantial improvement of the ATLAS laser pulse contrast inevitable.

Different schemes have been considered to improve the contrast on the short, picosecond time scale, including all optical techniques such as the implementation of a XPW or OPA stage (Sect. 3.1.1). However, these schemes require the implementation of an additional stretcher compressor pair (double CPA) and therefore did interfere considerably with the planned laser architecture of the upgraded system. Apart from these complications, the ability of these schemes to clean on very short time scales is questionable as they operate before the final re-compression and hence, are not able to correct for temporal side wings introduced by imperfect pulse re-compression. Hence, to ensure best contrast conditions for the envisioned thin foil experiments at MPQ, a re-collimating double plasma mirror system was designed for the ATLAS laser system.

Design and Engineering

The underlying concept of the ATLAS plasma mirror was to implement the pulse cleaning system as an integral part into the ATLAS laser. Thus, the new system should provide the option to clean the pulse right after the pulse compression, before sending it to any of the experimental chambers via the beamline system. This concept is different from the plasma mirror systems built in other laboratories [6, 7], which were directly attached to an experimental chamber and therefore could only serve one specific experiment. Due to the lack of space in the laser hall, it was decided to build the plasma mirror on top of the optical table of the laser—thereby making it a truly compact system.

However, this idea poses major challenges to the technical design of the plasma mirror. A highly confined space of $2 \times 1.5 \times 0.5$ m was allocated to the plasma mirror system, at a height of $\sim$2 m above the ATLAS laser. Installing heavy vacuum chambers at such a height is a major problem and therefore the weight of the whole system had to be taken into account for the technical design. Thus, instead of using one big vacuum vessel, the PM was assembled out of three small chambers, interconnected with rather light vacuum tubes to reduce the overall weight.

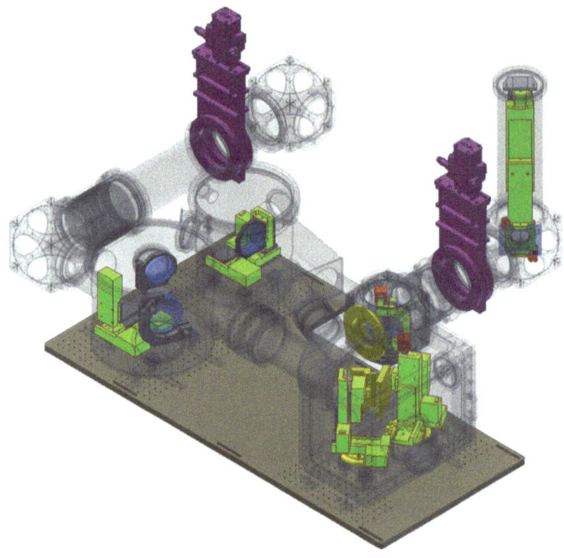

Fig. A.3 *ATLAS Plasma mirror: Engineering design.* The whole system was planned and designed with millimeter precision in advance to the mechanical construction

Moreover, as standard parts were too space consuming, almost all mechanical parts were custom made and adapted to the specific requirements. In order to have full control on the optical alignment in vacuum conditions, the optical system of the plasma mirror was fully motorized, comprising fifteen translation stages in combination with another fourteen tip-tilt mirror motorization units—making it a quite sophisticated experimental setup on its own. A snapshot of the three dimensional engineering drawing is shown in Fig. A.3.

Optical Setup and Pulse Characterization

The upgraded ATLAS laser system showed substantially worse laser pulse contrast than expected from the older system [10] and thus two consecutive PM reflections had to be used to reach contrast conditions sufficient for nanoscale targets. Two PM substrates were set up in the near field of the converging (expanding) beam at a distance of 15 mm (PM1) and 10 mm (PM2) with respect to the focus and an angle of incidence of 50°, as depicted in Fig. A.4. Taking into account day-to-day variations in the final output energy of the laser system, this setup corresponds to an estimated fluence of 90–120 J/cm^2 (200–270 J/cm^2) on the first (second) PM, in accordance with the optimal fluence values given in literature. To ensure high PM reflectivity, the polarization on the PM surface was set to s-polarized using a polarization rotating periscope that was implemented in the beamline system in front of the PM (Fig. A.4). The optical damage observed after each shot on the PM substrates was $\sim$3 mm in diameter, which in turn allowed for $\sim$150 shots by translating the PM surfaces before breaking the vacuum and swapping PM substrates. Having powerful beam

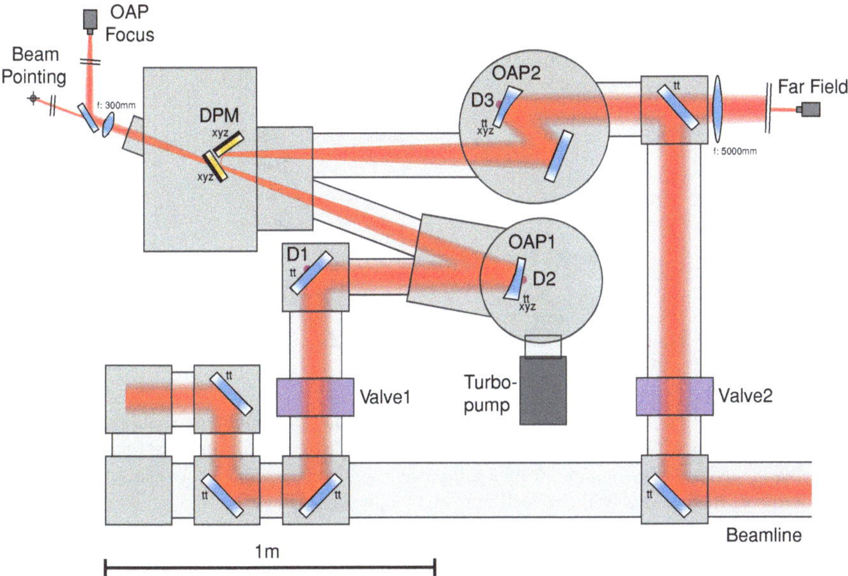

Fig. A.4 *ATLAS Plasma mirror optical setup—D1–D2*: diodes used to define input axis of the laser beam. *OAP Focus*: imaging to check focus quality and focus position. *Beam pointing*: cross-hair to check beam pointing of the focused beam. *D3*: diode to check beam position after PM reflection, defines in combination with *far field* the output axis of the beam. *Far field*: check re-collimation and output direction. All diodes were monitored with imaging cameras (not shown here). Labels: *xyz*: three-axis translation stage, *tt*: tip-tilt mirror motorization

alignment diagnostics is crucial for the routinely operation of such a sensitive optical system. Hence, a variety of different alignment marks were introduced to ensure stable operation of the system, most of them are schematically shown in Fig. A.4.

The energy transmission through the DPM system was monitored online by focusing the light leakage of a beamline mirror located next to the target chamber to a calibrated diode detector. Energy transmission values of ~40 % were typically observed from DPM shots. As it turns out, this value is a combination of the PM reflectivity and additional losses introduced by the beamline system. Here, the rotation of the laser polarization necessary to ensure high PM reflectivity gives rise to a slightly reduced reflectivity of the beamline mirrors. Hence, the reflectivity of the DPM system on it's own is expected to be rather close to 50 % (or 70 % for each PM reflection) in agreement with other DPM systems [6, 7].

The intensity distribution on target was examined carefully using the full power ATLAS beam and directly comparing bypass and DPM shots. Figure A.5 shows the focal distribution observed from a full power DPM shot, demonstrating excellent focusability and no degradation of the focal distribution as compared to the bypass shots. This observation is quite remarkable, taking into account that the reflections from two plasma surfaces as well as the precise alignment of three sensitive off-axis parabolic mirrors are involved in this experimental setup.

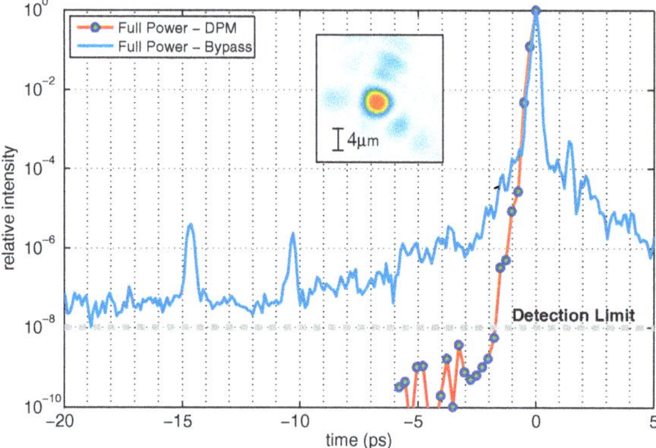

Fig. A.5 *ATLAS laser pulse contrast.* Third order autocorrelation (Amplitude Sequoia) measured at full power (running all amplifiers) with and without the double plasma mirror. *Inset* focal distribution measured in the ion target chamber from a full power double plasma mirror shot showing excellent focusability of the beam after the reflection from two plasma surfaces

To evaluate the contrast improvement of the DPM system, a scanning third order autocorrelation was carried out using the full power laser system (Fig. A.5). A remarkable contrast enhancement by at least three orders of magnitude can clearly be seen from the measured autocorrelation curves, which potentially could be increased even further using optimized anti-reflective PM coatings as well as an improved plasma debris shielding in between the PM substrates. The autocorrelation measurement suggests that ionization of the DLC targets takes place at around −2 ps, which in contrast would happen already many 10 s of picoseconds before the peak without the use of the DPM system. In agreement with that measurement, no ion signal could be observed from bypass shots, clearly demonstrating the key role of the designed DPM system for thin foil experiments carried out at the MPQ ATLAS laser system [11].

References

1. Doumy G, Quéré F, Gobert O, Perdrix M, Martin Ph, Audebert P, Gauthier JC, Geindre JP, Wittmann T (2004) Complete characterization of a plasma mirror for the production of high-contrast ultraintense laser pulses. Phys Rev E 69:026402
2. Dromey B, Kar S, Zepf M, Foster P (2004) The plasma mirror—a subpicosecond optical switch for ultrahigh power lasers. Rev Sci Instrum 75(3):645–649
3. Ziener Ch, Foster PS, Divall EJ, Hooker CJ, Hutchinson MHR, Langley AJ, Neely D (2003) Specular reflectivity of plasma mirrors as a function of intensity, pulse duration, and angle of incidence. J Appl Phys 93(1):768–770
4. Nomura Y, Veisz L, Schmid K, Wittmann T, Wild J, Krausz F (2007) Time-resolved reflectivity measurements on a plasma mirror with few-cycle laser pulses. New J Phys 9(1):9

5. Freidberg JP, Mitchell RW, Morse RL, Rudsinski LI (1972) Resonant absorption of laser light by plasma targets. Phys Rev Lett 28:795–799
6. Wittmann T, Geindre JP, Audebert P, Marjoribanks RS, Rousseau JP, Burgy F, Douillet D, Lefrou T, Ta Phuoc K, Chambaret JP (2006) Towards ultrahigh-contrast ultraintense laser pulses—complete characterization of a double plasma-mirror pulse cleaner. Rev Sci Instrum 77(8):083109
7. Lévy A, Ceccotti T, D'Oliveira P, Réau F, Perdrix M, Quéré F, Monot P, Bougeard M, Lagadec H, Martin P, Geindre J-P, Audebert P (2007) Double plasma mirror for ultrahigh temporal contrast ultraintense laser pulses. Opt Lett 32(3):310–312
8. Henig A, Kiefer D, Markey K, Gautier DC, Flippo KA, Letzring S, Johnson RP, Shimada T, Yin L, Albright BJ, Bowers KJ, Fernández JC, Rykovanov SG, Wu HC, Zepf M, Jung D, Liechtenstein VKh, Schreiber J, Habs D, Hegelich BM (2009) Enhanced laser-driven ion acceleration in the relativistic transparency regime. Phys Rev Lett 103(4):045002
9. Kaluza M, Schreiber J, Santala MIK, Tsakiris GD, Eidmann K, Meyer-ter Vehn J, Witte KJ (2004) Influence of the laser prepulse on proton acceleration in thin-foil experiments. Phys Rev Lett 93:045003
10. Hörlein R (2009) Investigation of the XUV emission from the interaction of intense femtosecond laser pulses with solid targets. PhD thesis, Ludwig–Maximilians–Universität München (LMU)
11. Bin J, Allinger K, Assmann W, Dollinger G, Drexler GA, Friedl AA, Habs D, Hilz P, Hoerlein R, Humble N, Karsch S, Khrennikov K, Kiefer D, Krausz F, Ma W, Michalski D, Molls M, Raith S, Reinhardt S, Roper B, Schmid TE, Tajima T, Wenz J, Zlobinskaya O, Schreiber J, Wilkens JJ (2012) A laser-driven nanosecond proton source for radiobiological studies. Appl Phys Lett 101(24):243701

Appendix B
Spectrometers

B.1 Wide Angle Electron Ion Spectrometer

Electron spectrometers typically resolve only a tiny fraction of the generated electron beams (solid angle of the spectrometer $\Omega \sim 10^{-6}$ sr) and thus do not provide any information on the spatial distribution. This is acceptable for well-known thermal distributions, which have rather low directionality and are uniformly distributed over large emission angles (tens of degrees, Sect. 4.5, Fig. 4.10). However, resolving the spatial distribution becomes important for highly directed electron beams pointing in a direction different from the laser axis (e.g. ponderomotive scattering, Sect. 2.2.5), or highly fragmented beams from laser plasma instabilities or electron beam filamentation.

To resolve particle beams over a broad angular range, a magnetic spectrometer with a large acceptance angle was designed and tested in experiment. The particle beam enters the spectrometer through an elongated slit, which is oriented perpendicular to the dispersion direction of the spectrometer. Particles ejected from the target at different angles enter the magnet at different positions, get deflected by the magnetic field and eventually hit the detector screen (scintillator).

The magnetic field distribution is deduced from the numerical simulation of the magnet geometry (CST) and re-scaled in magnitude to the actual field strength, which was determined from Hall probe measurements. As can be seen from the lineouts taken in longitudinal and transverse direction (Fig. B.1b, c), the magnetic field is strongly inhomogeneous, owing to the large separation of both magnets. Moreover, in longitudinal direction, the magnetic field leaks out of the magnet substantially. Thus, in order to shield the magnetic field in front of the yoke and avoid that any particle deflection takes place before the particles entering the spectrometer, the slit aperture was machined out of two (ferromagnetic) iron plates, which were directly attached to the yoke.

To deduce the electron energy from the recorded signal, collimated, monochromatic particle beams with different energies and propagation directions were tracked from the source to the detector. From that tracking, contour lines of constant energy

© Springer International Publishing Switzerland 2015

D. Kiefer, *Relativistic Electron Mirrors*, Springer Theses,
DOI 10.1007/978-3-319-07752-9

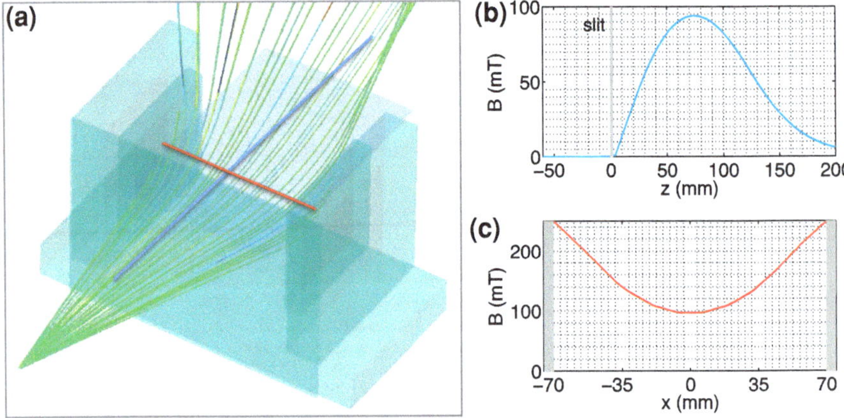

Fig. B.1 *Wide angle spectrometer simulations: magnetic field and particle tracking.* **a** Electron trajectories of multiple, monochromatic (5 MeV) electron beams, emitted in different directions. Longitudinal and transverse lineouts of the magnetic field distribution are shown in (**b**, **c**)

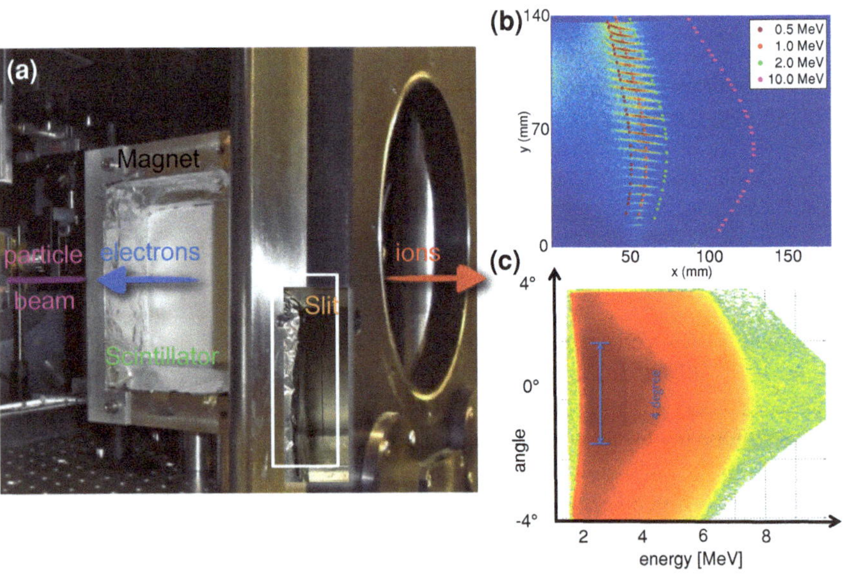

Fig. B.2 *Wide angle spectrometer experimental setup.* **a** spectrometer setup at the ATLAS ion chamber, **b** electron signal (using a pinhole array instead of an entrance slit), **c** proton signal

can be extracted. Figure B.2b shows the detector signal obtained from the interaction of the MBI laser pulse with a nanoscale foil, superimposed with the isoenergy lines extracted from the simulation. The shape of the measured signal perfectly matches to the contour lines of the simulation, demonstrating the proof-of-concept of the spectrometer.

Unlike the ion wide angle spectrometers designed very recently [1, 2], this spectrometer is capable of measuring both ion and electron distributions simultaneously within a large acceptance angle. It is an ideal tool to study the angular dependence of electrons accelerated from laser nanofoil interactions and investigate more sophisticated interaction schemes such as the cancelation of the transverse momentum of laser-accelerated electrons using a secondary reflector foil [3]. In-depth spectral analysis and experimental studies testing the idea of momentum switching will be part of future work.

B.2 Magnetic Field Measurements & Spectrometer Dispersion Curves

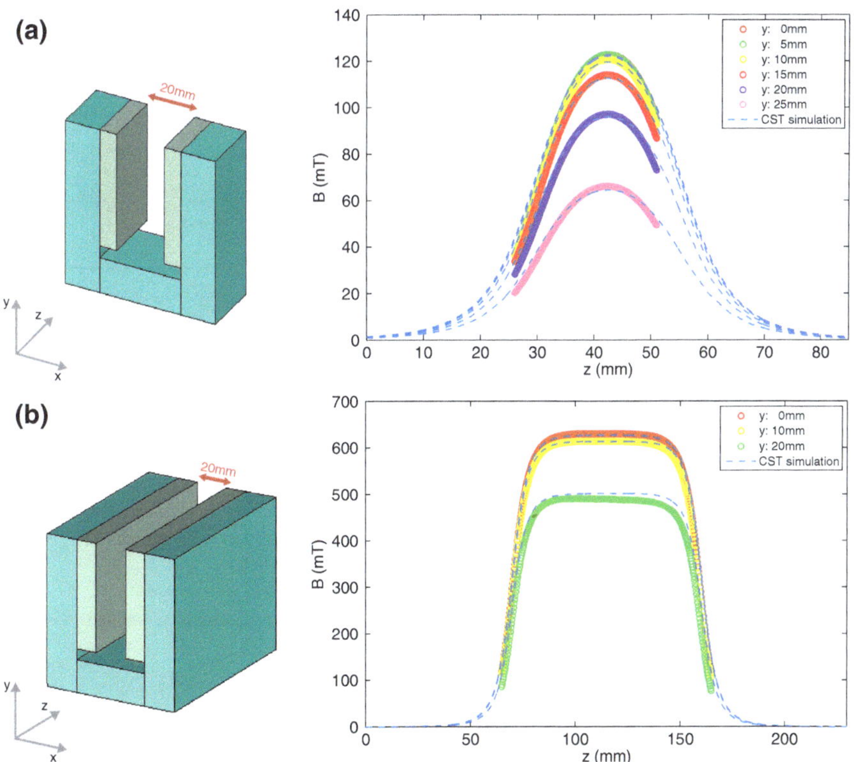

Fig. B.3 *Magnets used in the electron spectrometer.* The yoke dimensions are **a** 25 × 65 × 70 mm using a pair of hard ferrite magnets (25 × 7 × 50 mm), and **b** 90 × 80 × 70 mm in combination with NdFeB magnets (90 × 10 × 50 mm). The magnetic field was evaluated along the z-direction for different y positions and constant x position (*center* of the magnet) using a Hall probe (*colored curves*). The magnetic field was simulated numerically using CST, where the remanence of both magnets was adjusted to the measured field maximum. Corresponding lineouts taken from the simulation (*dashed lines*) show excellent agreement with the actual field distribution

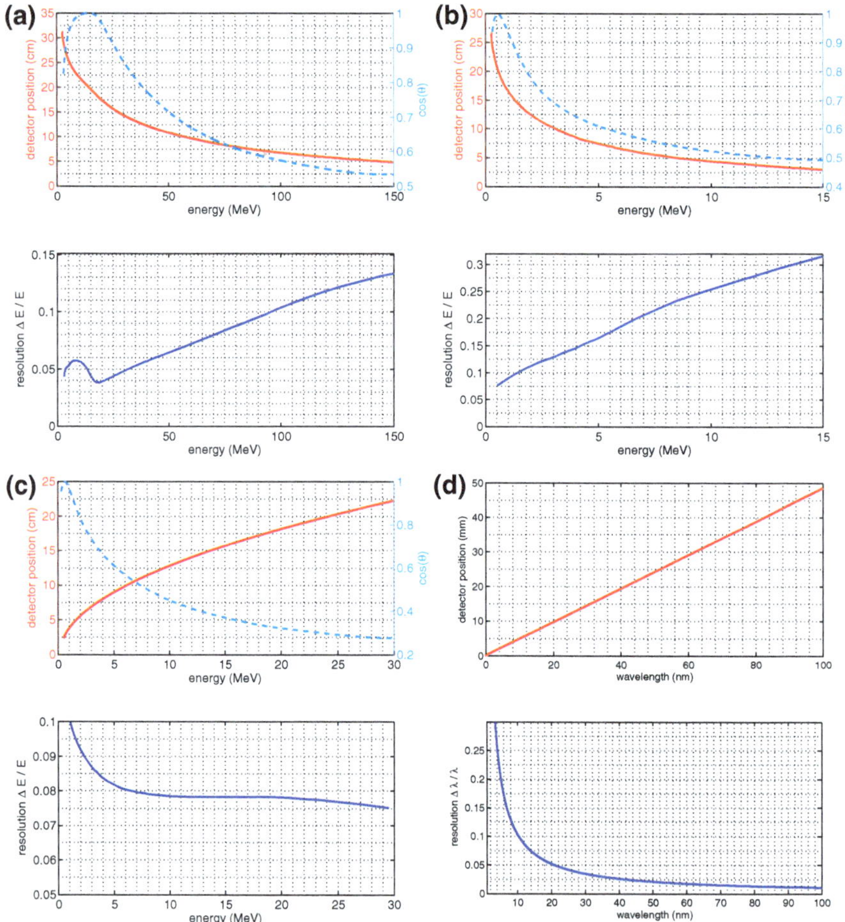

Fig. B.4 *Spectrometer dispersion curves and spectral resolution of the instruments employed in the experiments.* Instrument *curves* of the electron spectrometers used at **a** LANL, **b** MBI, **c** RAL and the XUV spectrometer used at RAL (**d**)

References

1. Chen H, Hazi AU, van Maren R, Chen SN, Fuchs J, Gauthier M, Le Pape S, Rygg JR, Shepherd R (2010) An imaging proton spectrometer for short-pulse laser plasma experiments. Rev Sci Instrum 81(10):10D314
2. Jung D, Horlein R, Gautier DC, Letzring S, Kiefer D, Allinger K, Albright BJ, Shah R, Palaniyappan S, Yin L, Fernandez JC, Habs D, Hegelich BM (2011) A novel high resolution ion wide angle spectrometer. Rev Sci Instrum 82(4):043301
3. Wu H-C, Meyer-ter Vehn J, Fernández J, Hegelich BM (2010) Uniform laser-driven relativistic electron layer for coherent Thomson scattering. Phys Rev Lett 104 (23):234801

Zeitfracht Medien GmbH
Ferdinand-Jühlke-Straße 7
99095 Erfurt, Deutschland
produktsicherheit@kolibri360.de